DATE DUE

Saving Nature's Legacy

Timothy J. Farnham

Saving Nature's Legacy

Origins of the Idea of Biological Diversity

Yale University Press
New Haven & London

Published with assistance from the foundation established in memory
of Philip Hamilton McMillan of the Class of 1894, Yale College.

Set in Simoncini Garamond type by The Composing Room of Michigan, Inc.
Printed in the United States of America.

Library of Congress Cataloging-in-Publication Data

Farnham, Timothy J.
   Saving nature's legacy : origins of the idea of biological diversity /
Timothy J. Farnham.
        p. cm.
   Includes bibliographical references.
   ISBN-13: 978-0-300-12005-9 (clothbound : alk. paper)

   1. Biodiversity—History.   I. Title.
   QH541.15.B56F37 2007
   577—dc22

                                    2006031296

A catalogue record for this book is available from the British Library.

The paper in this book meets the guidelines for permanence and durability
of the Committee on Production Guidelines for Book Longevity of the
Council on Library Resources.

10 9 8 7 6 5 4 3 2 1

To

Ellen, Anabelle, and Julia,
for support, for love, and for constantly reminding me
about the most important things in life

Contents

Acknowledgments

I would like to thank my mentors and friends at the Yale School of Forestry and Environmental Studies, where much of the research for this book was completed. Many thanks especially go to Steve Kellert for his past and present guidance. His advice always has the wonderful quality of being both supportive and challenging, and he continually pushed me to think carefully about the choices I needed to make in researching such an unwieldy topic as the history of the biodiversity concept. I would like to thank John Wargo and Steven Stoll, whose expert advice and keen observations helped me immensely, especially in the early stages of developing this topic. Many thanks also to all my friends in New Haven, who made my time there a memorable and intellectually stimulating experience.

I would like to thank my colleagues in the Department of Environmental Studies at the University of Nevada, Las Vegas—David Hassenzahl, Helen Neill, Krys Stave, and Patrick Drohan—for their encouragement and advice as I wrote this book.

Many thanks to Yale University Press and their fine editorial team, especially Jean Thomson Black, for publishing this book and making it so much better than it originally was as a manuscript. Thanks also to the anonymous reviewers, whose insightful commentaries undoubtedly helped to improve the quality of the book.

I received generous funding from several sources, without which I would not have been able to complete my research. My heartfelt thanks go to the Yale School of Forestry and Environmental Studies; to the Yale Institute for Biospheric Studies, for providing the G. Evelyn Hutchinson

Fellowship; and to the Heinz Foundation, which awarded me a full year of funding as a Teresa Heinz Scholar for Environmental Research.

I would also like to thank my parents, who have supported all my choices throughout my life. And I reserve my final thanks for Anabelle and Julia, who have been an endless source of joy and distraction, and for Ellen, whose energy and willingness to encourage and support me through the difficult times is constant and unwavering.

Abbreviations

AAAS	American Association for the Advancement of Science
AEC	Atomic Energy Commission
AOU	American Ornithological Union
ASTA	American Seed Trade Association
AWI	American Wildlife Institute
CBD	Convention on Biological Diversity
CE	Conservation of Ecosystems (US/IBP program)
CEQ	Council on Environmental Quality
CGIAR	Consultative Group on International Agricultural Research
CIMMYT	International Center for the Improvement of Maize and Wheat
CITES	Convention on International Trade in Endangered Species of Wild Fauna and Flora
CT	Conservation Terrestrial
ESA	Endangered Species Act
FAO	Food and Agriculture Organization
FCER	Federal Committee on Ecological Reserves
IARC	International Agricultural Research Centers
IBP	International Biological Programme
IBPGR	International Board of Plant Genetic Resources
ICSU	International Council of Scientific Unions
IRRI	International Rice Research Institute
IUBS	International Union of Biological Sciences
IUCN	International Union for the Conservation of Nature and Natural Resources
IUPN	International Union for the Protection of Nature
MAB	Man and the Biosphere Programme
MMPA	Marine Mammal Protection Act
NAS	National Academy of Sciences
NFMA	National Forest Management Act

NRC	National Research Council
NSSL	National Seed Storage Laboratory
OTA	Office of Technology Assessment
PARC	Predator and Rodent Control
SAF	Society of American Foresters
TNC	The Nature Conservancy
TVA	Tennessee Valley Authority
UNEP	United Nations Environment Programme
UNESCO	United Nations Educational, Scientific, and Cultural Organization
USAID	United States Agency for International Development
USDA	United States Department of Agriculture
WCS	World Conservation Strategy
WWF	World Wildlife Fund

Introduction

Biological diversity, for many of today's conservationists and environmentalists, may seem like a term that has been around for a long time. In the various arenas of environmental protection and management, the loss of biological diversity is a ubiquitous concern. It can be found on the agendas of national and international environmental organizations, in reports and documents written by government agencies, and in curricula of undergraduate and graduate programs in natural resources management and environmental studies. As a prominent conservation biologist wrote in 1991, "Biological diversity . . . is the most popular buzzword in conservation these days, and rightly so" (Noss 1991: 230). In the years since this comment, the term's pervasiveness arguably has elevated it beyond "buzzword" status. The maintenance of biological diversity is now firmly entrenched as a leading issue in the environmental community.

In the scope of conservation history, however, the term has only recently gained its broad popularity. Its presence in both academic and popular journals has grown dramatically since the early 1980s. In a database compiled by the Institute for Scientific Information (which currently indexes more than fifteen thousand peer-reviewed journals), a keyword search using the term *biological diversity* and its shortened form *biodiversity* turned up zero references in 1980 and 1981 and only 7 references for 1982. By 2005, that number had grown to 3,905 (figure 1). Published books about biological diversity, whose numbers have likewise grown considerably, show the breadth and complexity of the issues involved. Applications of the term range from economic valuations to ethical treatises, from scientific methodology of measurement to cultural and sociological implications of biodiversity loss. Individual books have focused on such specific topics as agricultural practices, biotechnology, ecotourism, intrinsic value, computer modeling, property rights, and global warming—all in the context of the conservation of biological diversity. Since the early 1990s, environmental organizations have devoted more time and resources to this issue as well, creating staff positions and programs, and government agencies involved in resource management have contracted reports and released studies examining activities and their impact on biological diversity. Legislation for protecting biodiversity has been introduced in Congress in several forms since the mid-1980s. From these trends and events, it seems clear that writers, readers, funders, scientists, environmentalists, and the U.S. government have increasingly found biological diversity to be an intriguing and important topic.

Biological diversity is more than simply a new name for *nature;* its definition requires a certain perception of the structure of the natural world. First, the central interest of biological diversity is the variety of all life on earth, from the smallest living organisms to the largest. Living things are primary. Any interest in geologic, chemical, and physical attributes of the natural world is placed in the context of their impact upon or connection to biological life. Second, biological diversity is most commonly considered at three distinct levels of organization: genetic diversity, species diversity, and ecosystem diversity. Other levels could certainly be used to describe life on earth, but these three have stood the test of time to become the accepted components of the standard meaning. The definition thus establishes a specific framework for how we are to consider the phenomenon of life and our impact upon it.

While the term carries the sheen of scientific authority, it is important to note that biological diversity—as we understand it in 2007—evolved chiefly in environmental circles. At its most basic level, biological diversity refers to the objective, measurable components that make up the variety of life, but the term also implicitly suggests concern over the human degradation of the environment. In short, when a person invokes biological diversity, it is more than likely that he or she is interested in discussing some facet of the environmental crisis surrounding the loss of natural variety. Early in its definitional history, it was clear that the term was being used to represent the ultimate source of the

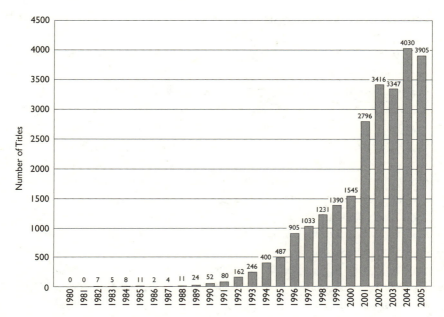

Figure 1. *Journal articles using the term* biological diversity *or* biodiversity *as a keyword, as identified in a database compiled by the Institute for Scientific Information. There were ten references previous to 1980 that used* biological diversity *in the abstract, title, or as a keyword, but none of the articles used the term as it would come to be defined in 1980 in its conservation context.* Source: *Scopus Database, Elsevier Science. Drawn by Bill Nelson.*

benefits that humans reap from a healthy, functioning environment. Not only did biological diversity provide a basic model for examining the natural world, it also was intimately tied, from its earliest usages, to the full range of conservation issues. Any activity that was damaging to life on earth—pollution, development, overexploitation, the growth of the human population—could be considered a threat to biological diversity and the benefits it supplied.

In this way, biological diversity can be characterized as a *concept* for conservation, and as such it represents the natural world as a repository of environmental values—the benefits that humans receive from nature. By calling biological diversity a concept, I do not mean to raise a philosophical issue over the "reality" of the nonhuman world; nor do I wish to debate the relationship between the human perception of nature and the "actual" nature that exists. Rather, I believe that in our attempts to ascertain what is most important to us, we need to construct models that purposefully highlight that which is valuable. By using these frameworks, we may work to protect the values that we have consciously or unconsciously identified.

The existence of such frameworks is particularly evident in conservation history. For example, it is easily shown that conservation concepts that came into existence before the rise of biological diversity emphasize the primacy of certain values. The concept of *wilderness* focuses on the numerous values—ecological, recreational, aesthetic— inherent in the protection of uninhabited, untrammeled land. *Resourcism,* a concept with roots in the conservation efforts of the early twentieth century, emphasizes the utilitarian value of commodities that can be derived from the natural world for human use. *Endangered species,* a concept with close ties to biological diversity, encourages us to consider not only the aesthetic and ethical value of protecting animals and plants close to extinction but also the potential future values that could be lost if any life forms disappear. These concepts, simply stated, are value-laden frameworks, containing specific ideas about what ought to be protected and why. The characteristic that distinguishes biological diversity from other conservation concepts is that it envelops a much larger segment of nature and encompasses a wider range of values.

Biological diversity has also come to be closely associated with science and the scientific value of the natural world. This is largely because a number of prominent biologists in the mid-1980s adopted biodiversity conservation as their cause. In 1986, the Society for Conservation Biology was founded as "a response . . . to the biological diversity crisis" (Soulé 1987: 4), and in the following years, a flurry of studies and articles were published with the intention of informing others about the facts surrounding the "current status of life on earth" (Wilson 1985: 20). This association with science has likely given biological diversity an advantage over the other concepts in conservation, for it surrounds the term with an appealing aura of objectivity and knowledge. While the other concepts are primarily connected to values that either are more subjective in quality or are economically based, biological diversity is perceived as representing a view of the natural world that is not only free of bias but also representative of the most accurate information that scientists can provide.

However, because the concept arose from discussions about the many benefits humans gain from nature and the danger of losing these benefits through environmental degradation, biological diversity encompasses much more than a target for scientific inquiry. It includes all of the values at stake in the maintenance of the living resources of the earth. Certainly, scientific value is one of these, as many biologists have made abundantly clear. Every species lost means, as one popular metaphor states, the tearing of a page from the book of life on earth and, subsequently, an irretrievable loss of evolutionary and biological knowledge. But scientific value is only one facet of biological diversity. The utilitarian economic value of the products of living resources is immense. The ecological value of the services that a healthy biosphere provides is practically beyond measure. The recreational, aesthetic, and ethical value of preserving genes, species, and ecosystems is also considered by many to be extremely high. In short, because of its breadth, biological diversity is able to subsume all of the values highlighted in other conservation concepts, presenting them as an integrated set.

Indeed, biological diversity has done more than simply co-opt other concepts' values; it has engulfed the concepts themselves. A primary reason for protecting wilderness is to maintain unspoiled natural areas for preserving biological diversity. The conservation of the innumerable economic resources we value is seen as a powerful supporting argument for maintaining diversity, ensuring the harvest of goods and services for many years to come. The plight of endangered species, while still a powerful conservation motivator in itself, is considered part of a larger strategy for protecting biodiversity, concerned not just with rare plants and animals but with all life. Biological diversity has become an umbrella under which various environmental interests can congregate and protect values collectively. For this reason, the concept has gained an unrivaled precedence in the conservation community.

Given the current pervasive quality of biological diversity in environmentalism, it is worth asking certain questions about its undeniable popularity. First, how did this ambitious concept for conservation evolve? In my research for this book, I began by tracing the definitional history of the term and reviewing influential ideas, publications, and events that contributed to its development as a conservation concept. It soon became apparent that biological diversity was able to gain favor among a wide variety of people because it succeeded in expressing a range of values and concerns that previously were scattered disparately throughout the scientific and conservation communities. It was also evident that the authors who proposed the first definition of the term in 1980 were consciously trying to formulate a concept that would bring various conservation constituencies together. In the years after the introduction of the term, it was appropriated by numerous groups, from those interested in genetic resources for agriculture to those concerned about the "unraveling" of impoverished ecosystems. The term struck a chord with those both inside and outside the traditional conservation community, and as it was used more extensively its conceptual framework began to take shape.

By 1987, the three-tiered definition identifying genes, species, and ecosystems as the hierarchical levels of biological diversity had established itself in the literature. This development inspires a second question: why were these three categories selected over others? One reason relates to biological diversity's implicit connection to environmental values. Simply stated, the three divisions of genetic, species, and ecosystem diversity rose above other possibilities because they best encapsulated the values gained from nature. For example, we value genetic diversity because of its economic importance in agriculture and pharmaceuticals, and because of its critical role in the perpetuation of species. We value ecosystem diversity because it provides a supporting medium in the production of various materials, ecological services, aesthetic resources, and species habitat. The significance of species diversity to humans is perhaps most immediately recognized, from the utilitarian value of species as sources of food and material to the aesthetic, psychological, and recreational benefits they provide for society. These types of values are often cited in the literature about biological diversity, even in publications where the concept was first being developed.

But even more fundamentally, the respective interests in genes, species, and ecosystems have clear histories in which the evolution of concern for their conservation plays a prominent role. For example, well before 1980 scientists mobilized to conserve genetic diversity for research and utilitarian purposes; hunters and fishers, humanitarians, and biologists sought the protection of certain species that they valued; and ecologists, hikers, and natural resource managers argued that preserving large tracts of land was essential. In order to understand the multiple sources of biological diversity as a concept for conservation, I examined the historical concern for each of these hierarchical levels in depth. Not surprisingly, I found that one theme common to each history is the gradual expansion of values exhibited in protection efforts over the years. As the number of values grew and as the values themselves became broader and more inclusive, the connections between the conservation of genes, species, and ecosystems became more apparent, providing a supportive medium for the rise of a single concept that could represent all three interests.

In addition, especially in the latter half of the twentieth century, there was a growing recognition of the value of diversity in general. Not only were conservationists and scientists expressing interest in the protection and generation of diversity in the natural world, but many outside of conservation had begun to value diversity in human society as well—diversity in schools, diversity in the workplace, diversity in the citizenry of the country. In all contexts, diversity was seen as a strengthening characteristic. As the country grappled with issues of race, gender, and religion, people worked to build inclusive communities and produce positive results in an inevitable and sometimes difficult integration process. As a consequence, diversity emerged as a normative good, and this cultural development likely contributed to the popularity of diversity in environmental circles.

There were numerous examples of the conservation community's specific interest in variety and diversity before 1980, many of which bore a special significance as precursors to the concept of biological diversity. Most prominently, it was apparent that among environmental interests, discussions of diversity often included statements about the wide range of values that we needed to protect. It was assumed that by maintaining diversity in the natural world we could keep all of our options open, preserving a greater variety of values by preserving the natural variety of the environment. In this way, we have come to believe that with greater diversity—whether cultural or biological—comes greater value. The existence of this general assumption supports the observation that biological diversity was influenced by its early association with environmental values. In fact, it is unlikely that the concept could have developed apart from them.

Because of this implicit connection with values and because of the significant role that the expansion of values has generally played in conservation history, I decided that it would be instructive to develop a vocabulary of environmental values that could be used in discussions about the evolution of concern for biological diversity. This vocabulary is not meant to be a strict model for research, but rather a helpful tool in distinguishing between different attitudes (or changes in attitude) that affected the recognition of certain benefits provided by the natural world. In addition, a brief discussion of environmental values seems appropriate, given their central importance to the modern concept of biological diversity. In proposing this vocabulary, however, I must make several qualifications. The list of values is by no means meant to be exhaustive. Rather, in reviewing the selected values typologies, I am simply trying to find values concepts that fit well with the historical trends in conservation. The purpose is to assist in the narration of a particular history. Also, although I provide definitions for the different values, in practice the values often overlap and share characteristics. The definitions should be considered only as a guide in deciphering the rather complex web of values associated with the conservation of biological diversity.

In attempting to trace the evolution of such a large concept, it was essential to establish boundaries for the research. This was a most difficult task. It would have been possible to look back over many centuries and across many cultures in exploring the human fascination and need for the living variety of nature. But because biological diversity was born out of a modern conservation mentality, I thought it would be best to keep within the traditional historical confines of contemporary environmentalism. For this reason, my research does not extend far back in time beyond the early twentieth century, nor does it reach beyond the literature of the Western, developed world. Biodiversity conservation has become an international movement, significantly affecting developing countries and the relationships between nations. But the concept arose primarily under the influence of the conservation communities in the United States and western Europe. The research presented here does include some analysis of international environmental organizations and their role in the evolution of biological diversity, but the history is largely told in the context of how biological diversity developed in American environmentalism. Given the international importance of biodiversity conservation today, I anticipate that some readers may believe this decision ensures a hopelessly lopsided narrative, one that reinforces the hegemony of the industrialized, developed world. But it is important to emphasize that the values inherent in the concept are representative of a Western perspective, and not necessarily of those in tropical countries where biological diversity is richest. As a conservation paradigm, biological diversity was born out of specific traditions, and to understand it we must narrow our focus to its most significant influences.

Finally, as in the previous sentence, I use the term *paradigm* to describe biological diversity throughout this book. To clarify, I do not mean to imply that other concepts have lost their efficacy in the environmental arena. Indeed, wilderness, resourcism, endangered species, and other concepts still remain powerful motivators for conservation. But the primacy of biological diversity as an anointed cause for contemporary environmentalism gives the concept a special status. Today it is difficult to discuss the conservation implications of any human activity without mentioning impacts on biological diversity. The concept has become pervasive in environmentalism's vocabulary. Certainly, it seems likely that biological diversity will one day fall out of favor and be replaced by a new paradigmatic term that will better embody the environmental values and needs of the times. But for the present, biological diversity maintains its position as the current paradigm for conservation efforts, a concept that provides a multilayered model for what we ought to protect in the natural world.

To conclude, I would like to quote Elliott Norse, who is credited (with Roger McManus) with publishing the first definition of biological diversity in 1980. In 1996, Norse wrote a brief history of the concept, using an analogy to describe how this paradigm evolved in conservation: "A movement in human society is like a river system with many beginnings, in which tiny drops of water form and coalesce within tiny catchments as rivulets that join within progressively larger catchments as small streams, larger creeks, and rivers that ultimately flow to the sea or into basins with no outlet, where they dry up. The analogy goes further: because catchments are of very diverse sizes and because precipitation is diverse among them, some catchments contribute much more to the river's flow and characteristics than others. Some rivers are readily visible at the surface; others flow underground, out of sight. . . . The concept of conserving biological diversity at three hierarchical levels . . . has many influents" (Norse 1996: 5–6). My goal in this book is to uncover the significant streams of conservation that flow into the larger river of the biodiversity movement. Norse's objective in his article was to "provide an on-the-ground perspective of the confluence of ideas that propels this River today" (Norse 1996: 5). In contrast, I seek an aerial view that takes in the broad expanse of the landscape and waterways that made the river possible. As Norse writes, "Undoubtedly, I will overlook some significant springs or tributaries" (1996: 5). The same apology applies to this work. But it is my hope that from the research presented herein, a reader can take away a map that reveals how certain influents (or influences) contributed to the ways in which we approach the conservation of the natural world today.

1 *Defining the Modern Concept of Biological Diversity*

What is biological diversity? Or, more appropriate for this discussion, how do people choose to define it? One characteristic that obscures the definitional history of the concept is that the two cognates—biology and diversity—are certainly familiar words, especially to individuals in science and conservation. Because of this, as one author points out, many writers and readers may assume that "everyone shares the same intuitive definition" (Gaston 1996: 1). Unfortunately, as familiar as the cognates may be, these base words are also quite broad in scope, so that, when they are placed together, the consequent lack of specificity makes the conceptual understanding of biological diversity rather difficult to pin down—particularly when it is left undefined in a text. For this reason, numerous authors of books and articles on biodiversity dedicate a short section to articulating what the concept encompasses.

The most common shorthand synonym is "the variety of life on earth," but recognizing that this general concept requires further elucidation, authors usually offer a more specific, formal definition. In a textbook entitled *Fundamentals of Conservation Biology,* Malcolm Hunter offered this assessment of the term: "Definitions of biodiversity usually go one step beyond the obvious—the diversity of life—and define biodiversity as the diversity of life in all its forms, and at all levels of organization. 'In all its forms' reminds us that biodiversity includes plants, invertebrate animals, fungi, bacteria, and other microorganisms, as well as the vertebrates that garner most of the attention. 'All levels of organization' indicates that biodiversity refers to the diversity of genes and ecosystems, as well as species diversity" (Hunter 1996: 19).

Here Hunter has spelled out two important qualities that character-ize most of the significant published definitions. First, biological diver-sity refers to *all* forms of life. While this point may seem obvious to conservationists today, it is important from a historical perspective, particularly when comparing biodiversity to past perceptions of the natural world. The implication is that if one is concerned with main-taining biological diversity, one cannot focus solely on certain cate-gories of life. Protecting game animals or endangered species is only part of a strategy for preserving biological diversity. To ignore the less noticeable or less charismatic microflora and fauna in assessing the value of an ecological system is to miss essential parts of its biodiversity. The focus of concern broadens from the favored components of na-ture—such as economically valuable, huntable, rare, or aesthetically pleasing species—to the entire range of life on earth.

Declaring such a large object of study, however, requires a system-atic approach to comprehending the great amount of information rep-resented by "life on earth." Thus, Hunter's second observation is that, in addition to encompassing all life forms, biological diversity extends over "all levels of organization" that humans perceive in the living world. He divides these levels into three hierarchical categories: ge-netic diversity, species diversity, and ecosystem diversity. This three-tiered definition (originally suggested by Norse et al. 1986) is consid-ered the most widely used characterization of the concept in the literature. By 1992, according to the World Conservation Monitoring Centre, it had "become widespread practice to define biodiversity in terms of genes, species, and ecosystems, corresponding to three funda-mental and hierarchically-related levels of biological organization" (Groombridge 1992: xiii).

It is surely significant that genes, species, and ecosystems have en-dured as the organizational schemata of choice. For the purposes of this book, the most important feature of these categories is that the three levels represent already established fields of study in conserva-tion. From the early agricultural interest in genetic or "germplasm" re-sources to the popular appeal of protecting endangered species to the holistic concern arising from ecological science for entire ecosystems, the idea of biological diversity provided an umbrella concept that united various conservation traditions. Because of this connection to historical concerns, the three-tiered definition was readily received by a broad collection of interests. Scientists from diverse fields, natural resource professionals, policy makers, agribusiness representatives, and all who played some role in conservation efforts were able to find common ground in terms of the three familiar levels. This linking of common concerns was originally intended to produce a shared vision of protecting nature and its resources. From the first published definition in 1980, the concept of biological diversity was designed to bring previously disparate interests together under the objective of maintaining the variety of life on earth. By tapping into already established conservation traditions, biological diversity had a potentially vast constituency that would fight for the protection of the living natural world.

It did take several years for the standard definition to be seen in terms of the three levels. In publications after the term's "formal" introduction in 1980, some authors proposed that other levels of organization be identified as composing biological diversity. For example, conservation biologists have often insisted that it is important to distinguish a level of population diversity between genetic and species diversity (Wilcox 1984; Soulé 1985). Other researchers have suggested that we ought to consider regional or landscape diversity above the level of ecosystem diversity (Noss 1983; Salwasser 1991). As two observers have pointed out, there is a certain flexibility in such an all-inclusive concept: "If biodiversity is defined as all of 'Life on Earth,' then one can describe biodiversity and list the elements of biodiversity on many levels. . . . By convention, most definitions list ecosystems, species, and genes as the elements of biodiversity, although one could just as easily list landscapes, populations, and alleles" (Perlman and Adelson 1997: 9). However, while these numerous other levels certainly could have "just as easily" been employed to classify diversity, none have been accepted into the popular definition. The triumvirate of genes, species, and ecosystems continues to be most commonly found throughout the literature.

An influential contingent in the conservation community emphasizes that in addition to these three categories, biological diversity is not only the actual pieces and assemblages of life on earth, but also the resultant processes and functions that those pieces, by interacting with one another, have set into motion. For example, here is a 1991 definition that came out of dialogues between conservation groups, government agencies, and industry, moderated by the Keystone Center: "Biodiversity is the variety of life *and its processes*. It includes the variety of living organisms, the genetic differences among them, the community and ecosystems in which they occur, *and the ecological and evolutionary processes that keep them functioning, yet ever-changing and adapting*" (Keystone Center 1991; emphasis added). To these conservationists, biological diversity is not simply a taxonomic filing system; it also includes all functional operations that occur among the components of the natural world.

The practice of adding *process* to the definition has inspired protest in some circles, particularly from those scientists who are trying to find practical ways of measuring biological diversity in order to use it as an indicator for management (Gaston 1996; Hawksworth 1995). Still others believe that the concept has become too diluted to be effective, and that attention has effectively shifted from what some view as the central task of slowing the extinction rate of species. But the popular understanding of biodiversity has seized upon the idea of including the preservation of ecological processes as one of the goals in the maintenance of biological diversity. In these terms, it follows logically that the structural conception that characterizes biodiversity—a nested hierarchy of organizational levels—essentially reveals a functional *interdependence* of all living things. The perception of the natural world as an interlocking network of mutually dependent parts has become commonplace for those who defend the protection of biodiversity. Significantly, just as the hierarchical levels have connections to past conservation traditions, so too does the idea of interdependence carry historical significance. The idea of nature as "the web of life" had existed for at least several decades before the popular environmental movement came into full force. Now, this older concept has found a new home in a more expansive conservation paradigm. The quality of interdependence was quickly absorbed into the biodiversity vocabulary. As the American Museum of Natural History put it on shopping bags from the museum store, "biodiversity" is now popularly presented as "the spectacular variety of life and the essential interdependence among living things."

Interestingly, the combination of a reductionist framework (the building blocks of the biosphere categorized into different hierarchical levels) and a holistic sensibility (ecological processes and functional interdependence) again makes biological diversity an appealing concept for a wide array of interests. From taxonomists to landscape ecologists, from citizens concerned about a specific species to those concerned with the climatic impacts of tropical deforestation, biological diversity has become a paradigm of nature conservation that all can rally behind. Certainly, there are those who believe that the expanding quality of the concept has rendered it useless as a conservation objective. As one critic writes, "Biodiversity is so all-inclusive that it has become meaningless" (Lautenschlager 1997: 679). But this very inclusiveness is what allows the concept to attract a significant amount of global attention and support. Exactly how much it has furthered the cause of conservation is open to debate. It is apparent that the concept has passed far beyond being a simple catchphrase to become the leading paradigm for nature conservation.

Before examining the historical roots of the concept in subsequent chapters, we will look at some significant moments in the definitional history of the term. When, for example, was the concept first introduced and defined? Who were the early proponents adopting the term and voicing concern over biodiversity losses? What events and publications were most significant in advancing the concept? Exploring the answers to these questions will not only help us to understand the specific path that biological diversity followed in its rise through the ranks of environmental concerns, it will also allow us to gain insight into the modern conservation mentality toward nature, and subsequently will provide a framework for an examination of historical sources of the concept.

The First Definition

The term *biological diversity* appeared sporadically in the literature before 1980, and in many of these instances, especially in the 1960s, the author used it to refer to a subject related to genetic or molecular differentiation within either individuals, populations, or species. For example, John Allen's *Nature of Biological Diversity,* a collection of papers published in 1963, was not a wide-ranging discussion about the variety of life on earth (as might be expected in today's environmental writing) but a work that "attempt[ed] to define factors responsible for the diversification of cellular structure and function" (J. Allen 1963: v). Several technical articles in microbiology published in the 1960s and early 1970s used the term biological diversity in a similar way.

One of the first usages of the term that hinted at its future role in conservation was in an article entitled "Experience with Pesticides and the Theory of Conservation," by N. W. Moore, published in the periodical *Biological Conservation* in 1969. Moore used the conservation problems caused by pesticides to introduce the need for an overarching goal in conservation. Moore declared in his abstract, "Conservation of diversity should be the primary aim of conservation," and he specifically called for protecting a "wide range of habitats in nature reserves" (N. Moore 1969: 201). While Moore mostly used the word *diversity* without a modifier, he did employ the term *biological diversity* twice in the article, once in reference to conserving habitats, and the other, significantly, when describing the objective of conservation: "They [conservationists] should be explicit that their fundamental aim is to conserve biological diversity either for its own sake or for the future good of mankind" (N. Moore 1969: 203).

Certainly, this was not the narrow conception of biological diversity as used in articles about cellular structure and function. Moore was using the term in an inclusive attempt to represent the valuable variety found in the natural world. But, as with all authors who would use the term over the next decade, Moore offered no formal definition. His intent, however, was evident, and the identification of diversity as valuable signifies an important trend in the conservation thinking of the time. Several other significant mentions of biological diversity occurred throughout the 1970s, but no author or group saw the need to publish a definition. As in Moore's article, the term was used as a general characterization for the variety of components of the natural world that were deserving of protection.

It was not until 1980 that the first definition of biological diversity was published. It appeared in the second chapter of the 1980 Council on Environmental Quality (CEQ) *Annual Report,* entitled "Ecology and Living Resources: Biological Diversity," written by Elliott Norse and Roger McManus. Norse was a marine ecologist who had studied at the University of Southern California and moved to Washington, DC, to work in policy circles. Over a twelve-year period, in addition to working for the CEQ under Jimmy Carter, Norse held positions with the U.S. Environmental Protection Agency, the Ecological Society of America, and the Wilderness Society. McManus had come to Washington from Duke University during the Nixon administration and worked in the Office of Endangered Species for the U.S. Fish and Wildlife Service, and subsequently moved on to work on wildlife policy for the CEQ. In the preparations for writing the 1980 *Annual Report,* CEQ chair James Gustave Speth had noted the growing concern in the global conservation community over the loss of animal and plant species and their habitats, and he had requested that the CEQ staff research the topic further (McManus 2006). Norse had just been hired at the time the report was being written, and he remembers that within his first two weeks on the job (in December 1979), a planning meeting was held that brought together the biologists on staff. Norse wrote in a 1996 article, "CEQ Senior Staff Member for Land Use and Wildlife Malcolm Baldwin mentioned that someone (Norman Myers) had been documenting destruction of tropical forests at rates heretofore unimagined, a finding with grave implications for their species. He asked me to write a new chapter for the next CEQ Annual Report on an unprecedented subject: the status of life on Earth" (Norse 1996: 6).

But because the charge of writing about the "status of life" was vague, Norse asked for clarification. It was apparent that his superiors were chiefly interested in species extinctions, but Norse suggested that the problem was larger. In addition to endangered species, there was a growing concern over what was commonly known as germplasm resources—the genetic diversity available for maintaining the health of agricultural crops. On a larger scale, "whole ecosystem types" were being degraded beyond repair: "We were talking about the loss of diversity at all stages," as Norse put it in an interview (Norse 1999). The term biological diversity was apparently not used in this first meeting, but Norse comments that he "realized that biological diversity is what included these other levels" (Norse 1999). As he wrote in the 1996 article, "Knowing no existing term that encompassed all that was being lost, we called it 'biological diversity'" (Norse 1996: 6).

In the CEQ chapter, Norse and McManus offered the following formal definition, identifying just two hierarchical levels of biological diversity: "Biological diversity includes two related concepts, genetic diversity and ecological diversity. Genetic diversity is the amount of genetic variability among individuals in a single species, whether the species exists as a single interbreeding group or as a number of populations, strains, breeds, races, or subspecies. Ecological diversity (species richness) is the number of species in a community of organisms. Both kinds of diversity are fundamental to the functioning of ecological systems" (Norse and McManus 1980: 32). The authors apparently decided to keep the conceptual scope of biological diversity fairly simple. Genetic diversity was conceived as the differences among individual organisms within a species, while ecological diversity—the equivalent of "species richness"—was a simple count of how many different species existed in an ecosystem. Norse remembers that he had originally planned on presenting the three different levels that are common in current definitions, but later decided with McManus that a diversity level above the species level—ecosystem diversity—was not logical, perhaps because there were no easily identifiable living units with which variety at this level could be measured and catalogued (Norse 1999). Species, it could be argued, were the components that determined the significant differences in ecosystems; thus ecological diversity could be best expressed in terms of the species that were present. However, it is noteworthy that while the ecosystem level was not part of the original definition, both "genetic" and "ecological" diversity were presented as "fundamental to the functioning of ecological systems." Still, Norse ultimately felt that this omission of a higher category of diversity was a mistake, and in later publications he revised the definition introduced in the CEQ chapter.

While Norse and McManus's chapter is important because it contains the first published definition, and while it is certainly a point of interest that the original definition had only two hierarchical levels rather than three, the more significant characteristic of the term's formal introduction was the context in which the authors discussed their "new" conservation concept. First, Norse and McManus clearly framed the crisis in terms of the wide array of benefits and values that humans would stand to lose if the degradation of diversity continued. They claimed that biological diversity was humanity's "greatest natural resource, on which we depend for food, oxygen, clean water, energy, building materials, clothes, medicines, psychological well-being, and countless other benefits" (Norse and McManus 1980: 31). Any person concerned with the conservation or protection of these "benefits" necessarily had to be concerned with the maintenance of biodiversity. The chapter included individual sections on material benefits, ecological benefits, food sources, energy sources, industrial chemicals, pharmaceuticals, and the psychological and philosophical reasons for preserving biological diversity.

In addition, because the concept was defined so broadly, the authors noted that biological diversity was threatened by a wide assortment of human activities. In discussing the threats to diversity, the chapter included sections on the impacts of human settlements, transportation, crop production, forestry, chemical pollution, overexploitation, and the introduction of exotic species—all of these presented in the context of the exponentially increasing human population. By identifying the numerous ways that biodiversity is valuable and the potential impact of human activities, Norse and McManus were suggesting that everyone had some stake and responsibility in the preservation of biological diversity.

This was, of course, a familiar argument in conservation circles. For years, environmentalists had been blaming civilization in general for multiple transgressions against the natural world. But no one had ever been successful in linking the concerns together with a mutual objective. Norse and McManus ostensibly solved this problem by naming a concern (diversity of life on earth) that was broad enough to include the full range of environmental values and a variety of environmental issues, while at the same time easily reducible into manageable parts. In short, through this introductory definition of biological diversity, Norse and McManus were trying to give all of conservation a unifying purpose that it could support communally. Most important, the CEQ chapter served to place this infant conservation term into the fertile medium of environmentalism's lexicon to see if it might grow into a useful concept for how to protect the natural world.

The Rise of Biological Diversity in the Language of Conservation

While a more detailed account of the term's use after 1980 appears in chapter 7 (following, chronologically, the respective histories on the concerns for species, genes, and ecosystems), a brief overview of the highlights in the development of the concept is provided here as introductory information.

The constituent interests that began to use the term after the publication of the CEQ chapter were strikingly varied. The first major event to focus on biological diversity by name—the U.S. Strategy Conference on Biological Diversity, which took place in November 1981—was co-sponsored by numerous government agencies and programs, including the U.S. Agency for International Development (USAID), the Department of State, the Department of Agriculture, and the Department of Commerce, and also more traditional environmental interests, such as the Department of the Interior, the Council on Environmental Quality, the Smithsonian Institution, the National Science Foundation, and the U.S. Man and the Biosphere Program. The range of interests is noteworthy and indicates that the link that Norse and McManus had made between biological diversity and the many resource benefits it supplies had caught the eye of the more strategic-minded policy makers in Washington. As written in the preface to the proceedings: "The long term maintenance of the earth's biological resource base has received growing attention in recent years. This has included concern about the survival of many plants, animals, and microbial species and the implications of a diminishing biological resource base for worldwide agriculture, public health, economic growth, and social development" (USAID 1982: iii).

While many of the conferees focused on the material benefits of genetic diversity in particular, others at the meeting noted the broad range of values inherent in protecting all living resources of the natural world, using the same kind of language and examples that Norse and McManus had employed in their work. In a short time, biological diversity had been pulled into the national and international policy spotlight, moving from the subject of a chapter in a government report to the central focus of a multi-agency conference, and it had attracted the attention of those government agencies not traditionally interested in environmental issues. It was an auspicious debut.

Prominent environmental writers began to use the term about this time, inserting it into the many discussions about the degradation of the earth's natural systems that were then popular. Two examples are Norman Myers and Paul Ehrlich. Myers is a British conservationist whose research in the 1970s on the degradation of tropical moist forests sounded the alarm that large areas of tropical ecosystems were quickly disappearing. He also published a book in 1979 called *The*

Sinking Ark: A New Look at the Problem of Disappearing Species, which enumerated in detail the many economic benefits that species provide for humans. It is evident that in writing the CEQ chapter Norse and McManus were strongly influenced by Myers's work; they cite *The Sinking Ark* nine times. Interestingly, Myers never used the term in the book. However, in his publications in the early 1980s, he integrated the use of biological diversity into his writing, which surely helped in the term's promotion and dissemination.

Paul Ehrlich, whose first claim to environmental fame was his 1968 work *The Population Bomb,* published a book with Anne Ehrlich called *Extinction: The Causes and Consequences of the Disappearance of Species* in 1981, in which he introduced his often-cited rivet-popping analogy: that is, each species is a rivet in an airplane in which the human race is traveling; with each species loss, the integrity of the plane decreases, until at some critical juncture it falls apart. Although Ehrlich does not use the term biological diversity anywhere in this book, in subsequent articles in the 1980s about extinctions, the protection of diversity becomes the key issue for him. In short, the fact that spokespeople like Myers and Ehrlich chose to frame their arguments in the context of protecting biological diversity was a telling step in the term's growth in popularity and an example of how it quickly made its way into more popular works of the times. In addition, Myers and Ehrlich continued the emphasis on biological diversity as the source of the many values and benefits that humans enjoy. Myers, in particular, focused on material benefits, but he also discussed at length the aesthetic and ethical values inherent in protecting biodiversity, and Ehrlich's emphasis on the ecological value of a healthy, intact natural world is characteristic of many of his writings.

Biological diversity's next major appearance in the policy world came in 1983, when Congress passed the International Environmental Protection Act. This legislation (among other purposes) amended the Foreign Assistance Act of 1961 by specifically directing the U.S. government to seek out assistance opportunities that would help developing countries to conserve their genetic resources. As part of the planning process to implement the amendment, federal agencies were given the task of drafting a "comprehensive government strategy for conserving biological diversity" in foreign nations, and the U.S. Agency for International Development, the same institution that had organized the 1981 conference, was directed to head a task force of representatives from eleven different agencies. The message in this choice of leadership was apparent: diversity was integrally connected to economic development, and the variety of agencies involved indicated that Congress believed that it was in everyone's interest to work for the conservation of biological diversity. With the help of a coalition of environmental groups led by the World Resources Institute, their *Interagency Task Force Report* came out in 1985.

Even with all this attention from the federal government, the issues surrounding the losses of biological diversity were still largely unknown to the general public. The event that served to popularize the concept was the 1986 National Forum on Biodiversity. The highly publicized convention "featured more than 60 leading biologists, economists, agricultural experts, philosophers, representatives of assistance and lending agencies, and other professionals," and attracted coverage from major newspapers and magazines (Wilson 1988a: v). This was the first time *biological diversity* was reduced to the more media-friendly *biodiversity,* and from the amount of press that the forum received, the contraction seemed to serve its purpose well.

One of the leading figures to emerge from the numerous experts involved in the convention was biologist Edward O. Wilson. Wilson was already a highly respected senior scientist at Harvard University known for decades of contributions to biological research. From his work with Robert MacArthur in the 1960s on island biogeography (see chapter 4) to his Pulitzer Prize–winning book, *On Human Nature* (1978), to his unparalleled expertise on ants, Wilson's ideas and words carried tremendous import with colleagues and media alike. Early in 1985, he published an article for *Issues in Science and Technology* that would be reprinted in the periodicals *Current* and *BioScience* and in a number of edited collections of essays. It would serve as the clarion call, preparing the media, the scientific community, and the conservation world for the upcoming National Forum and the arrival of biodiversity as a leading conservation issue. In this article, "The Biological Diversity Crisis: A Challenge to Science," Wilson laid out his eloquent argument for the scientific value of biological diversity and why the cataloguing and protection of all life on earth is a "mission worthy of the best effort of science" (Wilson 1985: 23). Wilson presented many facts and anecdotes that would become the standard reference points for anyone writing about biological diversity: he highlighted the work of entomologist Terry Erwin, whose cataloguing of insects in the canopies of tropical forests showed that humans had identified only a fraction of the species that exist; he listed several utilitarian and economic arguments from Norman Myers; he cited the dramatic figure of 24,500 square miles of tropical rain forests ("the size of West Virginia"; Wilson 1985: 25) lost each year. The article culminated in Wilson's grand systematics project, "a complete survey of life on Earth," and his call to recognize that "systematics is an indispensable handmaiden of other branches of research [and] . . . the fountainhead of discoveries and new ideas" (Wilson 1985: 27). This article was, in many ways, a unifying rallying cry for the scientific community, and a signal to those outside of conservation circles that biological diversity was set to become the next mission for both activist-minded scientists and the popular environmental movement.

The rise of the biodiversity crisis to national prominence was solidified when Wilson published the papers presented at the forum in a 1988 book entitled *Biodiversity*. Many viewed this work as seminal both to bringing the biodiversity crisis to popular attention and to securing its reputation as a scientifically supported environmental issue, particularly because it had a luminary like E. O. Wilson as its spokesperson.

The Three-Tiered Definition

With so much public attention, a succinct and clear definition of biological diversity, especially for nonexperts, was becoming increasingly important. The "variety of life on earth" was commonly used, but even the general public needed more specific qualities to grasp the meaning of the concept. Norse and McManus's definition of "genetic" and "ecological" diversity was still the standard, but as Norse recounts, he had long felt after the publication of the CEQ chapter that a distinction was needed to distinguish between species and a higher level of biological organization (Norse 1999). It was a logical division to make, likely because it paralleled the general levels that researchers used to look at ecological problems, and would further clarify the interests that should be concerned with biodiversity loss.

In 1986, Norse teamed up with several scientists and natural resource managers to publish *Conserving Biological Diversity in Our National Forests,* a book directed at laypeople who were interested in conservation activities on federal lands, particularly those managed by the U.S. Forest Service. Here, a three-tiered definition was first formally presented, spelled out simply for nonspecialists: "Biological diversity is the diversity of life. Ecologists studying the patterns and processes of life tend to focus on three levels of biological diversity: genetic, species and ecosystem diversity. The most familiar level, species diversity, varies greatly from place to place. . . . A lower, less obvious level of biological diversity is genetic diversity within species. . . . There is also a higher level of biological diversity. Different physical settings have more or less distinctive communities of species. This is ecosystem diversity" (Norse et al. 1986: 2–3).

It is significant that Norse and his colleagues framed the three levels as the functional tiers at which research is performed. Genes, species, and ecosystems, suggested the authors, are the divisions of diversity that best represent how scientists, economists, and others view the natural world and frame their inquiries. The authors' words seemed to express the way that people had studied life on earth over the preceding years. In this way, the three-tiered definition was not a novel development or a new concept forced into the conservation vocabulary. Instead, it is perhaps better characterized as the authors' observation of how certain sectors of society were examining and valuing the natural world. Likely, it is for this reason that the definition slipped easily into use and was readily accepted by a wide range of constituencies.

The convention of referring to genes, species, and ecosystems was firmly established when the Office of Technology Assessment (OTA) published its extensive and influential report *Technologies to Maintain Biological Diversity* in 1987. This document, along with the associated commissioned studies, became a popular source to cite in the growing number of articles and books that were published on the topic in the late 1980s and early 1990s. Its formal definition reaffirmed the three levels offered by Norse and his colleagues: "Biological diversity refers to the variety and variability among living organisms and the ecological complexes in which they occur. Diversity can be defined as the number of different items and their relative frequency. For biological diversity, these items are organized at many levels, ranging from complete ecosystems to the chemical structures that are the molecular basis of heredity. Thus, the term encompasses different ecosystems, species, genes, and their relative abundance" (OTA 1987: 3).

While we cannot be absolutely certain of the exact reasons why just three hierarchical levels were chosen and not more, several possibilities present themselves. The three-tiered definition, introduced when interest was expanding exponentially, likely provided an easily communicated framework that did not sacrifice the complexity inherent in the issues concerning biological diversity. But most important, the categories of genes, species, and ecosystems played into an already established network of conservation imperatives. These were terms with which both environmentalists and the general public were familiar. To unify the levels under an umbrella concept additionally reaffirmed the popular notion that all components of nature were interconnected. Biological diversity was providing the conceptual vehicle that could carry all of the disparate but related concerns and ideas about the conservation of the natural world.

Because the OTA report would be examined closely by a wide range of interests and would influence future publications, the genes-species-ecosystem triumvirate became etched into environmentalism's terminology. As the report itself noted, "The interest in maintaining biological diversity has created a common ground for a variety of groups concerned with implications of a reduction or ultimate loss of the planet's genetic, species, or ecosystem diversity" (OTA 1987: iii). One need not look any further for the "variety" of interested parties than the six congressional committees that supported the OTA report. As one article reviewing the study noted: "The Senate and House agriculture committees, for example, are concerned about the narrowing of genetic diversity for agriculture and other commercial enterprises. . . . The Senate foreign relations and House foreign affairs committees are worried that the loss of species and habitats in developing countries may have disastrous consequences for those nations' economic development. . . . Although most discussions of biological diversity focus on the tropics, the House Committee on Merchant Marine and Fisheries is concerned about the loss of species and habitats in the United States. The House Committee on Science, Space, and Technology is interested in the role U.S. science and technology can play in conserving biological diversity worldwide" (Shen 1987: 709). The three-tiered definition apparently aided the concept in fulfilling its promise as a link not only between conservation interests but between a wide range of federal social and economic interests as well. The three categories gave the term a tangible quality that helped it to advance in a public arena. As evidenced by the interest generated by the OTA report, the concept, which had been fully engaged by the scientific community the year before, was now readily accepted into the U.S. national policy circles.

Conclusion

The years between 1980 and 1988 were a crucial time in the development of biological diversity as a conservation issue: after the first formal definition in the 1980 CEQ report, the 1981 Strategy Conference lifted the biological diversity crisis into the world of national and international policy; writers like Myers and Ehrlich helped to spread the word among conservationists; the 1985 *Interagency Task Force Report* brought the issue to the attention of Congress; the 1986 National Forum mobilized the scientific and environmental communities; the three-tiered definition was introduced by Norse and his colleagues in 1986 and solidified by the 1987 OTA report; and, in 1988, with the publication of the National Forum papers in the book *Biodiversity* edited by E. O. Wilson, awareness of the crisis reached unprecedented heights. Thus, by the late 1980s, the concept of biological diversity had fully matured into the umbrella concern that Norse and McManus had proposed.

Certainly, there were still formative events and publications to come in the 1990s, among them the Convention on Biological Diversity (CBD), an international treaty that opened for signature in June 1992 at the United Nations Conference on Environment and Development in Rio de Janeiro. But the definition and broad foundation of the concept were established well before the Rio Earth Summit. Largely, the CBD is noteworthy for connecting the conservation of biological diversity to sustainable development and for highlighting the issue of equitable compensation for the use of any country's biological resources. As such, the treaty evolved into an agreement between the two distinctive macroeconomic regions of the globe: the developed, industrial northern countries, and the still developing, largely poverty-prone southern nations. For the north, it was evident that the primary goal of the treaty was the maintenance and protection of the biological resources upon which some of the developed world's wealth was based. For the south, the more important objectives were those that addressed issues of access, ownership, equity, and financial and technological aid for conservation of biological diversity.

But beyond the identification of the biodiversity crisis with these international issues, the CBD did not change the basic meaning of the concept. By 1992, the term biological diversity had grown from an esoteric, seldom-used, vague expression to an encompassing concept representing the concern for all life on earth at multiple levels of organization. Because its development was influenced by a range of constituencies, such as conservationists, ecologists, government agencies, agricultural interests, biological scientists, and industry, the environmental values associated with the protection of biological diversity were numerous. Undeniably, the foundation of the concept was built on the multiple utilitarian, ecological, and scientific values that characterized the early publications. But as more groups became interested in the concept, it was apparent that a broad variety of values—including recreational, aesthetic, humanistic, ethical, and cultural values—were at stake in the protection of biological diversity.

This complete representation of values was a major reason that the cause of protecting biological diversity became so popular. Not surprisingly, a defining characteristic of the concept's history has been that its development at different times was guided by certain influential parties that would adapt the concept to address the conservation issues perceived as most pressing. From the concern over the loss of species to worries about conserving genetic resources to efforts to protect ecosystems and ecological and evolutionary processes, the concept of biological diversity would carry all of these conservation issues under one banner. Such breadth caused some conflicts but ultimately served to bring together previously disconnected interests and attract the attention of important parties that had traditionally ignored the environmental agenda.

The following chapters will explore the sources of these concerns and the historical expression of certain environmental values in an attempt to further illuminate the context that enabled biological diversity to climb into prominence in conservation circles over the last two decades of the twentieth century. Although such an inquiry cannot claim to identify all sources, we can, by systematically examining the historical efforts to protect the three different levels represented in the definition, trace a confluence of values and concerns over time. It is evident that in earlier years there were efforts to protect specific kinds of species, genes, and ecosystems for a narrow set of values. By the second half of the twentieth century, however, the values had expanded, and a general concern for diversity and variety had surfaced in the attempts to protect all three hierarchical levels. Those levels are represented by three conservation traditions—concerns for species, genes, and ecosystems—whose separate histories can be investigated. But first, to guide us in examining those histories, we will look briefly at the terminology of specific values, or, as it has been called here, the vocabulary of environmental values.

2 A Vocabulary of Environmental Values

The concept of biological diversity refers to far more than objective, measurable natural phenomena. The environmental community has adopted it as one of the most effective frameworks for understanding why we need to protect the natural world and for creating links between environmental well-being and numerous social values. How did this framework arise? What were the social and political attitudes that laid the groundwork, such that the federal government, environmental organizations, the scientific community, academia, and the public ultimately embraced the concept?

To answer these questions, we will need a practical vocabulary for the ways in which humans value the natural world. By noting how our culture has expressed such values, we may better understand how these principles have influenced our efforts to protect nature or, in today's conservation language, to maintain biological diversity. "Values" in this context does not refer simply to economic worth, although that kind of value has been one incentive for protecting biological diversity. Rather, environmental values help to explain why humans have found some parts of the natural world worthy of protection and respect and other parts unworthy. *Value* has two principal meanings in this context: it can be "internal . . . an attitude in the mind of a person . . . [or] external . . . the worth of an object in relation to other objects" (Steinhoff et al. 1987: 34). Obviously, internal values (attitudes) and external values (measures of worth) affect one another, but it seems likely that attitudes first must form and, to some extent, become manifest before some measure of worth is assigned. Therefore, attitudinal values are the ones that are pertinent for this research, in that they can be used to characterize broader social and cultural trends. This chapter describes environmental values in the conservation community over the past century as a means for developing a working vocabulary for different historical attitudes and motivations. This will help us to articulate how the social expression of values toward nature has changed along with the rise of concern for biological diversity.

Numerous books have been written on the subject of environmental values, and more generally, the literature has widely debated the value of nature (Owings 1976). From the natural law theories of the Stoics to Christian theological discussion of nature's purposes to the natural philosophies of Enlightenment thought, Western society over the centuries has examined our place in the natural world (Petulla 1980b). In more recent literature, discussions of environmental values can be found in numerous disciplines, including history, philosophy, sociology, and various natural resource professions. Because the purpose here is to develop a vocabulary of values, works in these fields that propose definitional categories or some type of classification system are of particular interest.

Such classifications in the values literature can be divided in two ways. First, values categories may be proposed by a learned observer who has some claim to objectivity. This is often the case in historical or philosophical works, in which authors propose categories of value that fit with the narratives or ideas they wish to relate. Second, values may be distinguished through social scientific methods that employ quantitative analysis to examine survey data or content extracted from source materials. The information is often grouped using statistical methods that reveal similarities in characteristics, and value types are derived from the results. In either case, the objective is to provide appropriate levels of distinction between the varying qualities of value that humans perceive in their interactions with the natural world.

Categories can be broad and sweeping or narrow and detailed, according to the objectives of the investigator. Joseph Petulla's often-cited book *American Environmentalism: Values, Tactics, Priorities* (1980a) contains an example of a commonly used broad classification system. Petulla declares that "the dual and occasionally paradoxical values of environmentalism" originate from "three sources of environmental traditions . . . biocentric, ecologic, and economic" (Petulla

1980a: 18–19). Briefly, the biocentric tradition is exemplified by the transcendentalist writings of Emerson, Thoreau, and Muir, which express spiritual and ethical motives for protecting the natural world. The ecologic tradition is distinguished by a concern for nature as an interdependent system, in which the health of the whole benefits both human and nonhuman members of the natural community. The economic perspective represents our utilitarian impulse, whereby we see nature as a collection of resources that we must efficiently manage to achieve the "wisest" use.

Petulla does not structure his discussion of values any further; instead he describes the changing roles of values in conservation history as variations within the larger framework of the three environmental traditions. Many writers have found the distinction between biocentric, ecological, and economic values extremely useful for commenting on motivations to protect the environment (see Yaffee 1982; Bormann and Kellert 1991; DiSilvestro 1993; Eldredge 1998). However, for the purposes of this study, such descriptors are too broad in scope to capture the nuances of change in cultural attitudes since the start of the twentieth century.

Others have offered more detailed taxonomies of values as tools for distinguishing finer characteristics within the trends of thought that Petulla proposed. In particular, philosophers have engaged in an important debate about the value of nature, articulating and defending a number of concepts and descriptors for how environmental value might be expressed. A leader in this field is Holmes Rolston III (1981, 1986, 1988, 1994) who proposes numerous categories of values.

Rolston's thought has gone through several iterations; the values categories presented here are from his 1988 book, *Environmental Ethics: Duties to and Values in the Natural World.* As the title suggests, Rolston does not draw as clear a line as other scholars do between the human evaluator and the natural entities that are valued. He considers nature to be an active participant in the valuation process: "Asking about values *carried* by Nature will let us make an inventory of how nature is valuable to humans" (Rolston 1988: 3). Rolston's analysis here blurs the boundary between subjective value (human attitudes) and objective (or intrinsic) value in the natural world. In first introducing his typology in 1981, Rolston had written, "I propose here to examine the kinds of value that arise in association with nature, being founded in

physical and biological properties there" (Rolston 1981: 114). While
Rolston would likely admit that all of his values are anthropocentric in
conception, he does not fully explain their existence as a human intellectual creation. As such, Rolston's values categories are helpful in describing a complete range of qualities that characterize the human relationship to the natural world, from material and physical dependence
to emotional, psychological, and spiritual foundations. These values
categories and their meanings are as follows:

Life-support value. Humans depend on healthy, functioning global
and local ecosystems to provide "life-support" for our needs and the
needs of other living things we depend upon.

Economic value. The natural world provides us with raw materials
that we can manipulate, and through our labors we produce artifacts
with utilitarian value.

Recreational value. We enjoy the natural world in other ways besides
our utilitarian and life-support needs, either through engaging in
physical activity or, more passively, as observers.

Scientific value. We experience intellectual pleasure by examining the
natural world and its components, trying to figure out nature's patterns and processes.

Aesthetic value. Rolston compares the appreciation of nature to the
appreciation of art, wherein beauty gives the viewer "purity of vision."

Genetic diversity value. This value is largely utilitarian, in terms of its
role in agriculture, and therefore appears to contain elements of Rolston's economic and life-support values.

Historical value. Rolston identifies two kinds of historical value: the
first is *cultural,* the natural world's role in the evolution of particular
societies (i.e., the frontier in American history); the second is *natural*
(closely related to scientific value), our world's evolutionary past.

Cultural symbolization value. Cultures adopt elements of the natural
world to symbolize certain key characteristics particular to each one.
Nature provides us with an expressive medium.

Character-building value. The natural world provides us with an opportunity to exercise and hone personal qualities of value, and to develop our sense of identity.

Stability and spontaneity value. The natural world provides us with stable conditions that allow life to prosper and with dynamic mechanisms that spur continual renewal and adaptation.

Dialectical value. Humans experience the Other in nature both as a challenge that builds our sense of self (as in the "character-building value") and as an example of what we are not.

Life value. Although similar in title to "life-support value," this category acknowledges the intrinsic value of the life phenomenon. Without the environmental conditions of the planet Earth, life would not be possible, and recognizing the normative good of life in general is, Rolston believes, an essential part of valuing the natural world.

Religious value. Nature provides an opportunity to ask and explore questions that ultimately may be beyond human comprehension. This value is realized in our consideration of more "spiritual" concerns.

Rolston's multileveled understanding demonstrates how difficult it is to establish a vocabulary for environmental values that captures all the philosophical nuances. However, in terms of the broader historical question of why people seek to protect biological diversity, some of Rolston's values categories, while pertinent to exploring the human relationship to the natural world, are not useful for explaining cultural trends in the protection of nature. For example, two values that are more philosophical in conception—dialectical value and stability and spontaneity value—have not been significant motivators for protecting biological diversity. They certainly represent a particular kind of value that humans find in the natural world, but they do not help us to understand larger societal movements. In addition, some values might be seen as subsets of other values. For example, genetic diversity value is related to economic value, life-support value, and scientific value; character-building value might be seen as a type of recreational value; cultural symbolism value could be included under historical value. Thus, Rolston's typology (with its philosophical roots) would need to be modified to help us understand historical trends.

If we find Petulla's categories too broad and Rolston's too detailed, we might turn to a discipline that interacts more closely with the natural world than do the traditional academic perspectives. Natural resource management professions have been debating the value of parks, natural areas, wilderness, animals, and plants for many years. Much of this work is categorized as a branch of either economics or sociology, according to whether value is defined as the worth of an object (including its economic worth) or the attitude of the person or group that experiences some aspect of nature. In particular, wildlife values—both economic and social—have received attention in efforts to characterize how we benefit from our interactions with other living things.

A paper entitled "The Future of Wildlife in Forest Land Use" (1947), presented at the Twelfth North American Wildlife Conference, was one of the earliest attempts to systematically categorize the value of wildlife. The author, Ralph King, asserts that "the total economic value of our wildlife resource is the sum of its several values plus the worth of the several services it performs" (King 1947: 456). King's paper, which was of fundamental importance for subsequent studies in wildlife values, proposes six general values of wildlife:

> Commercial values—the income derived from the sale of wild animals or their products or from direct or controlled use of wild animals and their progeny. . . .
>
> Recreational values—moneys expended in the pursuit of wildlife in connection with sports and hobbies.
>
> Biological values—the worth of the services rendered man by wild animals, e.g., insect and rodent control, sanitation, suppression of diseases, conversion, tillage, fertilization, etc.
>
> Social values—the values accruing to the community as a result of the presence of wild animals.
>
> Esthetic values—the values of objects and places possessing beauty, affording inspiration and opportunities for communion, contributing to the arts through music, poetry, literature, and painting, and possessing historical and patriotic significance. . . .
>
> Scientific values—values realized through the use of wildlife as a means for investigating certain fundamental and widespread natural phenomena that may affect man's interests either directly or indirectly. (King 1947: 456–457)

King's typology has been "widely applied" and is considered one of the most "useful" frameworks "for classifying the broad range of wildlife values" (Steinhoff 1980: 12). However, while these categories seem to represent the appropriate level of detail, they still have certain drawbacks. First and most important, King's formulation of values was meant mainly to assess the total *economic* value of wildlife. This focus is certainly appropriate for many purposes in wildlife management, particularly for convincing policy makers and the public of the value of protecting wildlife. But as a guide for examining trends in historical values toward nature, an economically based classification is somewhat limited. More appropriate for historical research would be a typology of attitudes that describes the motivations of individuals or groups for preserving the natural world. Still, King's selection of such categories as "social values" and "esthetic values" (although he does not very clearly define the former) suggests that he was also looking beyond traditional venues of economic evaluation, perhaps intending to stimulate dialogue and further study, so that less economically obvious values might be recognized and somehow translated into useful, quantitative data.

Some important value concepts are missing, however. For example, there is no obvious category for *ethical* values in King's typology, and clearly ethical concerns have played a defining role in the history of conservation efforts. And no place is allowed for *humanitarian*-based sympathy toward animals, historically an important motivation for protecting the natural world (Thomas 1983). Finally, King leaves no room for discussion of dis-value, of *negative* feelings toward nature. Additional categories are needed, and the expanded perspective of an attitudinal-based typology is essential for a more complete values vocabulary that can be used for discussing historical cultural trends.

Social ecologist Stephen Kellert has proposed a values typology that improves on such limitations (1979, 1980a, 1980b, 1996, 1997). In contrast to many other discussions of environmental values, Kellert's research is empirical, consisting of extensive surveys of individuals carried out in the late 1970s. From these survey results, Kellert derived a typology of nine basic values based on a social-psychological conception of attitudes, as opposed to the economic criteria upon which King based his categories. Attitudes, as Kellert writes, refer to "broadly integrated feelings, beliefs and values possessed by individuals" (1980b: 31). Kellert worked with a large number of people and developed value categories using statistical methods. It is interesting to note that, despite his more rigorous scientific method, the nine categories that Kellert identified in the late 1970s have certain details in common with the categories discussed above. These definitions are taken from his 1996 book, *The Value of Life:*

Utilitarian: Practical and material exploitation of nature

Naturalistic: Direct experience and exploration of nature

Ecologistic/scientific: Systematic study of structure, function, and relationship in nature

Aesthetic: Physical appeal and beauty of nature

Symbolic: Use of nature for language and thought

Humanistic: Strong emotional attachment and "love"

Moralistic: Spiritual reverence and ethical concern for nature

Dominionistic: Mastery, physical control, dominance of nature

Negativistic: Fear, aversion, alienation from nature (Kellert 1996: 38)

Kellert's typology also works well as a tool for looking at larger trends in American culture. In later studies and publications, he expands the use of his typology to explain attitudinal values of particular demographic groups, including those defined by age, gender, race, and educational level (Kellert 1996). These analyses, while based on surveys of individuals, reveal the flexibility of his typology, which is easily adapted to describe the motivations of groups in contemporary society. Significantly, Kellert also studied changes in trends of values expression in American society by examining newspaper articles over the previous one-hundred-year period (Kellert and Westervelt 1981). Thus, the typology has already been employed in a historical context, characterizing the broader attitudes represented within a culture over a particular period of time. It would seem, then, that his typology would be applicable to a more qualitative historical inquiry.

Kellert says that each of his nine values has many different degrees of manifestation. Although the values are permanent, biologically based, components of the human mind and body, their *expression* within individuals can vary extensively; values can "atrophy" if they are not nourished or can be "overexpressed" if improperly encouraged or cultivated. Kellert's typology is therefore an excellent theoretical tool for examining changes in values (or more precisely, expressions of values) relating to historical and cultural trends. In this way, for example, we can discuss how early twentieth-century expressions of "aesthetic" value compare with more recent expressions of it. In addition, within Kellert's flexible framework values can expand and become more inclusive in their expression over time. This is an important characteristic in the present context, as it provides a useful device for describing the changes in values expression that have occurred in conservation.

Most important, Kellert's typology has much to offer as a vocabulary for discussing how humans value biological diversity. His book *The Value of Life* (1996) focuses directly on this topic, describing in depth the connections between the nine basic values and the human concern for biodiversity. While originally conceived as a tool for measuring attitudes toward animals, Kellert's typology has been resilient and adaptable when applied to more general aspects of the human relationship with the natural world. The values can be used to describe not only human attitudes toward individual animals, but also our attitudes toward entire species, taxa, natural communities, and ecosystems. Because Kellert's work coincides with the rise of the concept of biological diversity, we might conclude that his typology "co-evolved" with this conservation paradigm.

With Kellert's typology of nine values as my foundation, supplemented by the work of Petulla, Rolston, and King, I propose the following glossary of values for my historical inquiry. The definitions should be considered simply as characterizations designed to help the reader understand the basic qualities of a particular value. This typology is meant to be a heuristic device that facilitates the discussion of the historical evolution of concern for biological diversity.

Utilitarian value. This is a broad values category that Kellert defines as a recognition of the practical and material benefits that humans gain from nature. As Kellert points out, all of his values could be classified as "utilitarian" because in theory they all provide "practical" benefits when they are expressed in a "healthy" way. But in this instance the focus is on the material, physical *use* of nature by humans for their own benefit. Two subcategories for this value are King's commercial value and Petulla's and Rolston's economic value. These alternative names represent a specific kind of utilitarian value, one that focuses on the natural world as a collection of resources that can readily be converted into monetary remuneration. It is a value that often characterized early attitudes toward the natural world.

Ecological value. This value is derived from Petulla's ecological tradition of thought, Kellert's ecologistic/scientific value, King's biological values category, and Rolston's life-support value. An important trend in the twentieth century was the growing awareness of the interconnectedness of the natural world and of how human well-being depends upon a healthy-functioning biota. As more scientists and laypeople developed rudimentary understandings of ecological concepts, they saw that the protection of the natural world was imperative because of the ripple effects that would occur if one component of the "web of life" were damaged or destroyed. Recognition of nature's role in pollination, pest control, predator-prey relationships, water purification, and global climate mediation encouraged efforts to minimize human impacts on ecosystems. Human interests and nonhuman interests are linked under this value, which emphasizes the interdependence of all living things.

Scientific value. As proposed by Kellert, King, and Rolston, this value emphasizes both the worth of the information collected and the intellectual satisfaction of systematically examining the natural world. While King identified the economic advantages that we would glean from scientific studies of nature, Kellert and Rolston discussed the pleasure that we feel when we strive to understand natural phenomena. Thus, this value is both utilitarian and social-psychological.

Aesthetic value. This value also was proposed by each of the three studies. King's definition is representative of how the descriptor *aesthetic* is employed in many natural resource management circles. It has come to mean any value that does not fit neatly into other economic categories—not only beauty, but also spirituality, artistic sensibility, and cultural or historical significance. Rolston's discussion emphasizes the "purity of vision" required to see beyond the pragmatic necessities of human use and experience the "natural essence" of the nonhuman world (Rolston 1988: 81–82). Taking Kellert's lead, we will define aesthetic value simply as the pleasure and emotion inspired by the beauty and splendor of nature. Other categories described below serve to distinguish values often grouped under the title of aesthetics.

Recreational value. While King focused on "sports and hobbies," Rolston emphasized the more contemplative activity of experiencing nature as an appreciative observer. In either case, recreation was understood as any human activity in the natural world undertaken solely for the pleasure of it. Most commonly, in the vast amount of literature that examines this value two kinds of recreation are distinguished: *consumptive* (e.g., hunting, fishing) and *nonconsumptive* (e.g., bird-watching, hiking). But because varied and often divergent attitudes are associated with recreational activities, some further distinction of values is necessary. Here, Kellert's typology is most useful. As one facet of recreational value, Kellert writes that "the dominionistic experience of nature can sharpen mental and physical competence through testing various abilities and capacities" (Kellert 1996: 20). The *dominionistic value* thus represents the desire of humans to employ nature as a place for taking on challenges and opportunities for exercising mastery. It applies equally to a hunter and a rock climber. Rolston's character-building value is closely related to this value. In a different expression of recreational value, the *naturalistic value* is the pleasure derived from exploring and discovering nature's complexity and variety. Again, both a consumptive recreator and a nonconsumptive recreator may enjoy this value, but in contrast to the dominionistic category, the naturalistic value emphasizes the joy of discovery and the awareness of the natural world.

Spiritual and ethical value. This value, the reverence and moral concern for nature and its components, closely parallels Petulla's biocentric tradition and largely derives from Kellert's moralistic value. While King seems to ignore the ethical value qualities that humans find in the natural world (although he did leave room for spiritual communion in his aesthetics category), the ethicist Rolston includes the theme of ethical commitment and reverence throughout his typology. Kellert's general definition seems most appropriate for categorizing and tracing the moral impulse in conservation history.

Cultural value. Rolston describes his cultural symbolism value as the qualities or objects in nature that have come "to express the values of the culture superimposed on [them], entering our sense of belongingness and identity" (Rolston 1988: 186), and provides as examples America's national symbol, the bald eagle, and South Dakota's state flower, the pasqueflower. Kellert's symbolic value is similar in conception but more broadly includes the value of the natural world in human mental development and language acquisition, a value that has not played a prominent role in the history of conservation. King may have included the concept of cultural value in his aesthetic or social values, but he does not define it as clearly as does Rolston. In short, Rolston's more specific definition best accounts for certain historical motivations for the conservation of nature.

Humanistic value. Present only in Kellert's typology, this value describes a powerful motivation for nature conservation, particularly at the beginning of the twentieth century. The humanistic value represents the sympathy for living nature (animals in particular), engendered by emotional attachment and a sense of kinship. As conservation biologist Archie Carr commented at the first U.S. Strategy Conference on Biological Diversity in 1981, "I would like to submit that one important reason for preserving species is because people like them" (USAID 1982: 59). Such feelings are easily discounted as merely sentimental, but their role in the history of nature protection cannot be ignored.

Negativistic value. From Kellert, the negativistic value (or "disvalue") is characterized by fear, hatred, and hostility toward natural entities. Kellert gives examples of animals or places that elicit a negative attitude: snakes, spiders, large carnivores, stagnant swamps, dark caves. In the twentieth century, the negativistic value was most evident in the common practice of distinguishing between "good" species and "bad" species, and specifically in the federally directed campaign against large predators, particularly the wolf, mountain lion, grizzly bear, and coyote (Dunlap 1988). While obviously not a positive episode in efforts to protect nature, the extermination efforts did provide a motivation and a focal point for those who voiced opposition to the indiscriminate killing of certain species.

In the following chapters, these brief descriptions of human values will provide a foundation for describing how Western attitudes toward the various levels of biological diversity have transformed since 1900, how the concept of biological diversity has evolved, and why it already had such fertile ground for growth when it was introduced in 1980.

3 The Concern for Species Diversity
The Tradition of Wildlife Conservation in the United States

A species, claims E. O. Wilson in his 1992 book *Diversity of Life,* is "the fundamental unit" of the earth's biological diversity (Wilson 1992: 35). Those with a more reductionist viewpoint might claim the gene's primacy, since it is the carrier of all pertinent information, while those with a holistic sensibility might insist on the singular importance of intact ecosystems for supporting all lower levels of life. Certainly, while it is difficult, if not impossible, to separate out the relative importance of genetic, species, and ecosystem diversity, it does seem that the species level has sparked the most interest, likely because it is the most basic unit that the human eye perceives in the nonhuman world.

More accurately, humans do not generally interact with species but rather with a particular species' individuals or groups of individuals. Very rarely can we appreciate the entirety of a species; when we do, it is often because the species has been so reduced that its viability is in danger. However, it is the concept of species that not only governs conservation efforts but also pervades our thoughts about the nonhuman world, for in our habits of classification we look to categorize the individuals that we regularly see according to their kind. Hence, when we speak of a gray squirrel, or an Indian tiger, or a furbish lousewort, we know what type of living thing we are talking about. These individuals show common traits with other individuals of the same species; by giving them a name, they become more familiar and our human inclination to place them in a category is satisfied.

There is some debate in the scientific community over the strict definition of a species. One of the more widely accepted definitions is from Ernst Mayr (1982), who adheres to the "biological species concept." "A species is a reproductive community of populations (reproductively isolated from others) that occupies a specific niche in nature" (Mayr 1982: 273). Arguments concerning this particular definition usually revolve around how to categorize naturally occurring hybrids and organisms that reproduce asexually. But scientific debates aside, from a cultural perspective there can be little disagreement over the influence of species as a classification mechanism that humans use to frame interactions with other organisms, either as individuals or as groups.

In particular, *animal* species have long been objects of human fascination. As C. A. W. Guggisberg writes in his 1970 book, *Man and Wildlife*, "Animals have been Man's longest and most fundamental preoccupation" (1970: 15). Although animals are only one component in the multilayered formulation of the concept of biological diversity, they represent the part of the natural world for which humans have the deepest affinity, revealed in many different arenas of human society, from work to recreation. Humans have depended upon animals—for food, labor, companionship, and the fulfillment of numerous psychological and cultural values—for hundreds of thousands, if not millions, of years. The relationship between humans and animals has gone through numerous changes, influenced largely by the evolution of human cultures and the attendant evolution of human attitudes toward animals (Thomas 1983).

This is not to deny that many plant species have also played significant roles in the development of human society, and that the value of members of the plant world is ostensibly equal to that of the animal kingdom. In fact, plants, by their unique ability to convert energy from the sun into digestible, nourishing sugars, fats, and proteins, are the foundation of all higher living organisms on earth. However, plant species have been more commonly recognized as carriers of valuable genetic material and as structural components of ecosystems. Human concern for plant species simply did not contain the same attachment to the individual as did concern for animal species. As a result, efforts to protect certain conspicuous animals dominated conservation efforts on behalf of species in the twentieth century. In addition, more so than invertebrates, insects, and microscopic forms of life, larger vertebrate animals are perceived by humans as the charismatic inhabitants of the nonhuman world and thus will continue to command the attention of human society. Because they are the part of biological diversity of which we are most aware, it is logical that animal endangerment would elicit the first historical expressions of concern. Indeed, federal protection of wildlife at the beginning of the twentieth century in the United States represented the first modern cohesive attempt in the United States (at a national level) to preserve the multiple values that humans find in the living natural world.

It is not my intention to offer a separate, exhaustive account of wildlife conservation in the twentieth century. Rather, I wish to show how concerns for wildlife were important precursors to concerns for biological diversity. In particular, the history of wildlife conservation will elucidate trends in environmental values within Western society. The dominance of utilitarian values progressively weakened as a wider array of values—recreational, humanistic, scientific, ecological, aesthetic, cultural, and ethical—grew more prevalent and came to characterize societal attitudes toward animal species. Accordingly, in a parallel development, the accepted definition of wildlife broadened from its original meaning (virtually synonymous with *game*) to include animals that represented a much more diverse set of values to humans. In addition, by the time the Endangered Species Acts of 1966, 1969, and 1973 were passed, it was recognized nearly universally that habitat was an essential component of animal species preservation, and that plant species, as integral, valuable parts of natural communities, ought to be afforded the same protection as animal species. In this way, the story of wildlife conservation reveals an important foundation for the rise of concern for biological diversity.

Animals as Products: The Early Dominance of Commercial Value

The first instances of concern for animal species in the New World by European settlers were almost invariably connected to preserving the food supply that wild game provided. As early as 1646, a closed season on deer hunting was established in Portsmouth, Rhode Island, in an effort to protect the species for future exploitation. This law provided a model for numerous others passed in the colonies, reserving a particular time of year from hunting and levying a fine on anyone caught violating the ordinance. Ostensibly, the values behind such laws were strictly utilitarian and economic: deer provided important commercial products for the colonies and served as a basic food source for the expanding Euro-American population. The early deer laws, however, were generally not enforced. Responsibility for arrests was passed down to local law authorities, and even though there were sometimes financial rewards for officers to apprehend offenders, convictions were rare. It seems that for the most part the natural abundance that greeted the first colonists provided little initial incentive to protect wildlife. Deer were perceived as a resource in which all could share, and there seemed to be no logical reason for state authority to restrict a man from securing food for his family. Likely both citizens and law officers believed this, and that explains why the early protection laws were generally failures. Ironically, efforts to conserve the utilitarian value of the deer were foiled because the realization of the value required the consumption of the animal and those receiving the benefits did not see any sense in curtailing their profits. Such is the case for many economically valuable resources in conservation history (Trefethen 1975; Lund 1980).

Deer were not the only target of the colonists. The diverse wildlife of North America provided many economic opportunities, from the lucrative fur trade to the sale of all kinds of wild meat. Trappers sought out beaver, otter, and mink, while market hunters—those who hunted and sold game for profit—bagged turkeys, grouse, quail, and numerous species of waterfowl. As the human population in North America grew, the demand for meat encouraged hunters to kill any animal that was saleable for its flesh. In contrast to deer, there were no early restrictions on popular game birds. As turkey and other popular targets became scarce in populated areas, the market hunters set their sights on the species that provided the most opportunity for large kills: migratory waterfowl. Huge numbers of ducks, geese, and other birds traveled in the spring and autumn between their northern summer breeding grounds and their southern winter havens. Driven by economic incentive, market hunters developed clever methods of attracting and killing the birds. Soon the firearm technology improved, with more consistent, rapidly reloading, and effective guns at the hunters' disposal. By the 1800s, the demand for wild game by markets, hotels, and the railroads had risen considerably. The pay for a successful day's worth of hunting was significantly more than what one could earn in a factory (Trefethen 1975; Matthiessen 1987).

It was not long before the profligate harvest of the wildlife resource began to take an obvious toll on numerous species. The heath hen, for example, a docile bird that was ill-adapted to avoiding hunters' guns, originally ranged from Maine to Virginia, inhabiting grasslands and barrens in states along the coast and on islands. In 1791, the decline of the bird's population on Long Island prompted New York State to institute a closed season and impose a fine for transgressions. By the mid-1880s, nevertheless, the species had vanished from most of its original range, surviving only on the island of Martha's Vineyard, Massachusetts, where the last individual heath hen died in 1932 (Matthiessen 1987: 67–68).

A more dramatic example, often cited in histories of wildlife protection and familiar to many outside of traditional conservation circles, is the fate of the passenger pigeon. Observers in the eighteenth and early nineteenth centuries described flocks of pigeons that blackened the skies and extended for miles. In 1806, ornithologist Alexander Wilson estimated a flock that he saw in Kentucky to consist of over two billion birds stretching for forty miles. John J. Audubon reported in 1813 that he had traveled through a roosting area many miles long in which he commonly heard limbs of trees breaking under the weight of the birds. The enormousness of the flocks made for easy hunting; a single shot would kill several birds in the massive flights. But market hunters eventually discovered that it was far more lucrative to locate the birds' roosts and use nets to achieve maximum kills. As late as 1878, netters were bagging huge numbers of birds. But by the last two decades of the century, the passenger pigeon population had plummeted. The species needed the dense numbers in order to reproduce successfully, and with the dramatic reductions the birds' mating behaviors were fatally disrupted. The demise of a species once so numerous was unthinkable, and the consequences of unrestricted hunting attracted the attention of many people. The last passenger pigeon died in captivity in 1914, and the story of the species' demise has become something of a conservation fable (Reiger 1975; Trefethen 1975; Matthiessen 1987).

But perhaps the most familiar conservation fable is that of the Amer-

ican bison, or buffalo, as the species is more commonly called. Estimated at sixty to a hundred million individuals, the bison herds ranged
widely on the North American continent before the arrival of Europeans. The intensive slaughter of the species between 1860 and 1880
took place for numerous reasons. First, the commercial value of buffalo products was high, and the expansion of the railroads allowed easy
mass transportation for the meat and hides that were taken. Buffalo
tongue was considered a special delicacy, so that, after killing the animals, hunters would simply remove the tongue and leave the carcass to
rot. Undoubtedly this practice hastened the species' decline: after the
few seconds it took to remove the bison's tongue, hunters were ready
to shoot the next target. Militarily, the removal of the buffalo herds was
a strategy to divest the American Indian peoples of their main source of
food. Even when it was evident that the species was being lost at a
frightening rate, no protection was allowed. In fact, a bill passed in
1874 by Congress, "to prevent the useless slaughter of Buffaloes within
the Territories of the United States," was pocket vetoed by President
Grant, likely because Grant and his advisors knew that the demise of
the bison would mean the demise of the Indian, leaving the West open
for settlement (Dunlap 1988; Goddard 1995).

By the last two decades of the century, only a few small herds were
left, and although the species was not driven to extinction, its dramatic
decline certainly helped to inspire future efforts in wildlife conservation. The sympathetic few at the time who knew the story of the buffalo were appalled at the destructive capability of the market hunters.
Naturalists wrote eloquent eulogies about the great herds that once
roamed the plains, decrying the greed and waste that so quickly decimated them. Thus, while other species like the heath hen and the passenger pigeon provided supporting examples, the buffalo slaughter
became a symbol for the destructive capabilities of human hunters. In
addition, in the latter decades of the nineteenth century, Americans
were sensing that the "closing of the frontier" was at hand, and for
some the end of the buffalo stirred nostalgia for a way of life. Thus, the
buffalo became associated with a cultural value that competed with the
commercial ones that largely caused its destruction. The species would
eventually become linked to a national heritage that sang longingly for
a "home where the buffalo roam." The buffalo would come to represent the freedom and wildness that the frontier had provided. Such

value, to those looking back, far outweighed the short-term profits enjoyed by the market hunters.

The Rise of the Sportsman

While other game species did not inspire the same sort of connection
to American cultural heritage as did the buffalo, another value was
gaining strength in the late nineteenth century, one that would come to
compete directly with the utilitarian commercial value placed on many
animals. With the new wealth brought by the industrial revolution,
more people began to have leisure time. Freed from the necessity of
hunting for their dinner, men revived the old English tradition of hunting for recreational enjoyment. Sportsmen's clubs began to appear
throughout the eastern United States, the first founded in 1844 in New
York (Cart 1971: 10). These groups quickly perceived that in order to
enjoy their hunting they needed to take steps to protect the game. As
early as 1848, the New York Sportsmen's Club had drafted a model law
for closed seasons on woodcock, ruffed grouse, quail, trout, and deer
in order to halt the decline in the populations of their favorite wildlife
targets (Trefethen 1975: 73).

These "gentlemen" hunters looked with disdain upon those who
hunted for money, and they decried year-round hunting and the killing
of large numbers of animals. Since the new interest in hunting for
recreation was backed by the money of its adherents, sporting articles
in popular periodicals and even entire magazines dedicated to the cel-
ebration of the sport became a fixture in the nineteenth-century press.
One of the most widely read writers was Henry William Herbert, who
published under the pen name Frank Forester. His rhapsodic descriptions of a day in the woods and his chiding concern over vanishing
game were popular with sportsmen of the day, and he surely influenced
many to work for restrictions directed at the market hunters. In fact,
the 1848 model law proposed by the New York Sportsman's Club was
based on a petition to the New York State government that had been
written by Herbert (Reiger 1975).

To some extent, the battle between the sportsmen and the market
hunters was viewed as a class conflict, pitting wealthy urban industrial-
ists against poor rural folk who depended upon the sale of game for
their livelihood (Warren 1997). The sportsmen obviously had more re-

sources than their less financially and politically able opponents. As the number of sportsmen's clubs grew, their influence on local and state governments increased proportionately. Using the voice of the sporting press, they argued that game was no longer essential as a food source, since agriculture could meet all of the growing nation's needs. This argument was voiced in its starkest terms in 1894, when George Bird Grinnell published an editorial in the popular magazine *Forest and Stream,* calling for the elimination of the sale of game in all seasons (Warren 1997).

It is important to note the difference in motivations of the market hunters and the sportsmen. The market hunters were products of a demand for meat by the growing nation's increasingly urban population. As opposed to the earlier times of settlement, many people from the middle class had less opportunity and desire to hunt for themselves but more resources to pay for someone else to procure their food. Although there were undoubtedly market hunters who enjoyed their work, the overwhelming value placed on wildlife species was economically based. Interestingly, the same social forces that encouraged the market hunters to kill large amounts of saleable game also brought forth the more recreation-minded sportsmen. Here were upper-class men who had achieved wealth through investment or business but who still desired the thrill of the hunt. This probably represented both dominionistic and naturalistic values—the sportsman motivated to exercise mastery over his prey and to enjoy more generally the invigorating experience of being outdoors. One might argue that the sportsman had the luxury to cultivate this recreational value of wildlife, having secured a plentiful income and bountiful larder. The sportsman's wealth and influence no doubt allowed him to advance his values over those of the market hunter. But the evident depletion of game species was also a major cause of the market hunter's demise. The market hunters had put themselves out of business by tragically overexploiting the resources.

Scientists and Humanitarians: Concern for Birds and the Lacey Act of 1900

Game species were not the only wildlife of concern in the late nineteenth century. It was, in fact, the widespread interest in protecting nongame birds that would play an important role in the passage of the first significant federal law protecting wildlife in the United States, the Lacey Act of 1900. While sportsmen and "meat" hunters were largely uninterested in many species of birds not deemed worthy as game, another industry saw new economic value for birds previously ignored by commerce. In the 1880s, the millinery trade began a fashion trend that would last for the next two decades. Extravagant hats with long head and tail plumes, breast feathers, and sometimes whole birds became an essential component of every woman's wardrobe. The lucrative industry paid large sums for feathers from species such as great white herons, roseate spoonbills, and egrets, and smaller amounts for gulls, terns, and some songbirds. The impact on these nongame populations was devastating; much like the buffalo, the birds' carcasses were often left to rot in the field after the desired feathers had been removed by the hunter (Cart 1971: 30–31).

Leading the charge against the plume trade were two groups who, along with the growing ranks of sportsmen, worked against the market and plume hunters through editorials, education, and lobbying for legislation. The first of these groups was made up by those interested in protecting species for their scientific value (Cart 1971). Ironically, the scientific naturalists of the nineteenth century had likely contributed to the thinning of nongame bird populations. From the early 1800s, specimen collection was considered an essential activity of every amateur or professional ornithologist. Fervent taxonomists would kill birds and gather eggs and nests to satisfy their curiosity. Often egg collections were displayed much like sport hunters exhibited the mounted heads of their most impressive kills. A large collection was seen as evidence of prodigious accomplishment. But professional ornithologists prided themselves on taking only what was needed in the name of science, denouncing the overzealousness of amateur naturalists. When the activities of the plume hunters became well known, the scientific community mobilized to stop the indiscriminate killing of their objects of study.

In 1883, several prominent zoologists founded the American Ornithologists Union (AOU). Modeled after its British counterpart, the AOU had general goals of advancing ornithology in North America and revising current lists and classification systems. Within a year, a special committee was formed for the protection of North American birds. The committee members worked for the next fifteen years to reach out to the general public and foster sentiment for the plight of nongame birds. In 1885, on the AOU's urging, Congress established the Division of Economic Ornithology and Mammalogy. C. Hart Merriam, a prominent AOU member, was selected as the head scientist. The division was housed in the Department of Agriculture and served as the precursor to the Bureau of Biological Survey. As its curious name implies, economic ornithology focused on the utilitarian and ecological (as in "economy of nature") value of nongame birds, mostly by presenting scientific evidence that nongame birds played crucial roles in controlling insect and weed populations. By casting nongame birds as our "feathered friends" in nature, Merriam and other naturalists used their scientific authority to show that such birds did indeed provide economically valuable services.

The scientists also advanced their cause on the legal front. In 1886, the AOU protection committee published the Model Law for nongame bird protection, directed at state legislatures. The proposed wording prohibited killing "any wild bird other than a game-bird" or taking "the nest or the eggs" of such a bird unless it was certified that the collection was being made for scientific purposes. By the turn of the century, nearly all states had passed some kind of law protecting nongame birds. Thirteen states used the wording proposed in the Model Law; twenty-three states identified "insectivorous," "song," or "harmless" birds as deserving special protection. The committee and the Division of Economic Ornithology and Mammalogy had done their publicity work well (Cart 1971).

Joining the scientific naturalists in their efforts against the plume and market hunters were sympathetic laypeople representing the humanitarian impulse toward animals. Popular interest in protecting nongame birds was high at the end of the nineteenth century, as indicated by the enthusiastic reception of an organization proposed by George Bird Grinnell in 1886. Grinnell, a charter AOU member and the editor of the popular sport periodical *Forest and Stream,* suggested calling the new group the Audubon Society, the purpose of which was to protect wild birds and their eggs. The response was so overwhelming (fifty thousand members in two years) that Grinnell had to give up the idea in 1889 because he had neither the time nor the staff to run a national organization. The Audubon Society was later revived in 1896 in Massachusetts (Fox 1981: 152).

But the initial response to Grinnell's group revealed the humanistic value that people saw in protecting the natural world, and birds in particular. Just as the sportsmen arose out of the wealth and urbanization of the newly industrialized country, so too did the sentimental nature lover. Popular literature began to reflect this inclination. John Burroughs's nature stories were widely read and enjoyed. Olive Thorne Miller wrote specifically with birds as her subjects; her titles included *Bird-Ways* (1883), *Little Brothers of the Air* (1892), and *Upon the Tree-tops* (1897) (Cart 1971: 129). Miller and other nature writers also contributed extensively to the popular magazines of the day. In these stories and essays, the animal protagonists were personified with intentions and emotions (Mighetto 1991), almost as if they were human. To kill them, in these terms, would practically be the equivalent of killing another human being. This humanistic value was intensified by reports of the profligate killing of the market and plume hunters. In combination with the men of science (who provided the more objective explanations for wildlife preservation), the nature lover/humanitarian was a potent force for protectionist policies and sentiments.

Aesthetic value, too, had an undeniable influence on the overall movement to protect both game and nongame animals. Sportsmen appreciated the beauty of their quarry and the experience of the wild; scientific researchers were surely motivated in part by the aesthetic pleasure of examining animals' forms; and nature lovers or humanitarians enjoyed communing with fellow beings of the nonhuman world. As Grinnell and Sheldon would later note, "An increasing number of people were interested in wildlife protection because these objects were beautiful to look at and ought to be preserved so that our successors may have the pleasure of seeing them" (Grinnell and Sheldon 1925: 201). Thus the economic and commercial values of the market hunters were slowly giving way to other values. This is not to say that such utilitarian values were completely overtaken. But with the sportsman, the scientist, and the humanitarian working to protect animal species, other values were certainly gaining in strength.

The combination of the sportsmen's opposition to the sale of game, the scientists' position on the beneficial effects of birds for agriculture, and the nature lovers' sympathetic moralism created a political groundswell that culminated in the passage of the Lacey Act of 1900, the first significant federal legislation related to wildlife management

(Cart 1971). Chiefly designed to support state wildlife laws by preventing transportation of illegally killed animals over state lines, the Lacey Act represented a careful first step by the federal government into what had previously been the states' jurisdiction. But the language of the law implied that bird populations needed human assistance, an aspect that state laws had not addressed: "The object and purpose of this Act is to aid in the restoration of such [game and other wild] birds in those parts of the United States . . . where the same have become scarce or extinct" (quoted in Cart 1971: 192). Although no direct protection was mandated in the law, the Lacey Act set the precedent of federal interest in the preservation of wildlife resources. The laws that followed enlarged federal jurisdiction and protected a progressively wider range of values.

Roosevelt, Antihunting Sentiment, and the Expansion of Aesthetic Value

Before other federal laws, however, came Theodore Roosevelt. Well known in conservation history as the "naturalist" president, Roosevelt aggressively pursued policies and actions that broadened the powers of the federal government and the executive office. In 1903, Roosevelt set aside what would become the first unit of the National Wildlife Refuge System by issuing an executive order: a federal bird refuge on Pelican Island was "reserved and set apart for use of the Department of Agriculture as a preserve and breeding ground for native birds." Technically, Congress had only authorized the president to designate lands as forest reserves, under the Forest Reserves Act of 1891. But Roosevelt needed no official permission; by the end of his first term, he had established fifty-one wildlife preserves, spanning seventeen states and the territories of Alaska, Hawaii, and Puerto Rico. In 1906, Congress passed the Antiquities Act, which gave the president the power to set aside lands of historical, archaeological, or scenic interest, and within weeks also passed a law protecting birds on federal lands "set aside as breeding grounds . . . by Executive Order." Although not an entirely clear mandate, Congress legitimized in this way Roosevelt's presidential decrees. By the end of his two terms in office, Roosevelt had left a land legacy for reserves, including those set aside for the National Forest System, of over two hundred million acres (Trefethen 1975: 124–125).

An avid sportsman, Roosevelt served as the most visible representative of hunters who believed in respecting and admiring the animals they hunted. During his years as president he inspired many Americans to take up the gun, "get back to nature," and experience the revitalizing tonic of the natural world. But the growth of interest in recreational hunting was also met by protest from a segment of the population that objected, on ethical grounds, to killing animals for sport. One agitator whose name would become closely tied to the anti-hunting movement was William Temple Hornaday. Hornaday, a taxidermist by training who had become director of the New York Zoological Park in 1896, was a charismatic personality whose zealous campaign to "stop the killing" was based on the charge that hunting was the single greatest cause of the decline in wildlife in the United States. His 1913 publication *Our Vanishing Wild Life* railed against the apathy of the "true sportsman, the government, the scientific community, and the public." In fact, there were few (even those who might have been considered his allies) who escaped his scathing criticism and vigorous attacks. He called for an "awakening" to "the Cause," and warned of the extinction of certain species if laws were not changed. "We are weary of witnessing the greed, selfishness and cruelty of civilized man toward the wild creatures of the earth. We are sick of tales of slaughter and pictures of carnage. It is time for a sweeping Reformation, and that is precisely what we now demand" (Hornaday 1913a: x).

The religious language was indicative of the energy and fanaticism that Hornaday brought to his diatribes. But as extreme as it may seem to us, Hornaday found an eager audience for his words. Those "nature lovers" who had joined with the scientists and sportsmen to pass the Lacey Act were beginning to see more conflicts than commonalities with the recreational hunting interests. *Our Vanishing Wild Life* was one of the first books dedicated to the cause of endangered wild animals. Hornaday used the book and his position at the New York Zoo to raise money for what he called the Permanent Wild Life Protection Fund, in essence an endowed source that Hornaday could use to travel, lecture, and write about his crusade. The fund reached its endowment goal of one hundred thousand dollars in only a few years, with contributions from such prominent persons as Henry Ford, Andrew Carnegie, and photographer George Eastman (Fox 1981: 149). Hornaday would go on to write several more books, including *Thirty Years War for Wildlife* (1913b), and would become easily the most visible campaigner for the antihunting movement.

In general, however, antihunting sentiment was too easily characterized by those in power as an emotional response of laypeople who did not understand the issues. The most effective arguments for protecting animal species were based on scientifically informed, utilitarian, and nonsentimental values. For example, in a 1915 article in the popular periodical the *Outlook* Roosevelt spoke highly of Hornaday's book *Wild Life Conservation* but also made light of those "foolish creatures" who protested hunting. Anticipating later concepts of wildlife management, Roosevelt argued that "when genuinely protected, birds and mammals increase so rapidly that it becomes imperative to kill them. If, under such circumstances, their numbers are not kept down by legitimate hunting . . . it would be necessary to have them completely exterminated by paid butchers" (Roosevelt 1915: 161).

In addition to Roosevelt's sporting management perspective, scientists continued to present their objective reasons for wildlife protection. During the final stages of the Migratory Bird Treaty in 1918, T. Gilbert Pearson, the president of the National Association of Audubon Societies, reminded the public of the importance of birds for insect control. In his article "To Conserve Food in America, We Must Preserve the Wild Life," published in the *Touchstone,* he not only advocated maintaining the breeding stock of game birds but also noted "the enormously valuable insect-eating" service that birds provide for the nation (T. Pearson 1918: 259). There were also those in the scientific community, continuing the tradition of the naturalists, who worked to pass the Lacey Act, calling for the preservation of wildlife because of its intrinsic scientific interest. Willard G. Van Name published an article in *Science,* entitled "Zoological Aims and Opportunities," which attempted to bring attention to the "protective work which is very important to science. . . . This is the protection of what remains of the unique and peculiar forms of animal and plant life that inhabit many of the remote islands . . . in various parts of the world" (Van Name 1919: 83). Although Van Name recognized that "the general public [could not] be expected to appreciate" the importance of preserving wildlife for scientific study, he believed he could motivate his fellow scientists to act.

As Roosevelt pointed out in his article in the *Outlook,* the aesthetic value of nature was another major reason why the general populace wanted to protect birds and animals. While multiple "conservation" programs that sought to use the resources efficiently ("the greatest good for the greatest number of people") were the hallmark of Roosevelt's presidential years, by bringing in the notion of aesthetics Roosevelt was suggesting a different kind of utility. The simple fact was that people found pleasure in looking at wildlife. "Man began to appreciate the need of preserving wild life, not only because it was useful, but also because it was beautiful." The interests of "unborn generations" were now considered as a reason for preservation: "Now there is a considerable body of public opinion in favor of keeping for our children's children, as a priceless heritage, all the delicate beauty of the lesser and all the burly majesty of the mightier forms of wild life" (Roosevelt 1915: 159). Here was an aesthetic argument that appealed to the masses and emphasized the naturalistic value of "getting back to nature" by pursuing one's quarry. Those who could not appreciate nature's creatures were depicted as "barbaric," while those who took action "to restrain the senseless destruction" were described as "enlightened" and "civilized." And although Roosevelt certainly never voiced support for the antihunting forces, he did provide a platform, a forceful aesthetic argument, upon which those concerned with wildlife preservation could find common ground and join forces. He attempted—and to a large extent succeeded—to find a popular reason for protecting the nation's birds and mammals.

Negative Value: The Eradication of Predators

While the aesthetic value and other positive contributions of certain animals promoted support for their preservation in the first quarter of the twentieth century, a different kind of value was forming toward other members of the natural world. As the American population settled the western half of the continent, ranchers and wool growers laid claim to wide expanses of country to raise their stock. Predatory animals soon became the target of an extermination campaign to protect the herds of domestic animals that now populated the region. The stockmen appealed to the federal government for help and the Department of Agriculture obliged. Ironically, it was the old Division of Economic Ornithology and Mammalogy, renamed the Bureau of Biological Survey in 1905, that would sponsor the new predator control program. The same office that had promoted bird protection was now managing the systematic killing of wolves, cougars, coyotes, eagles, bears, and any other animals that were seen as a threat.

Although the federal government had been helping to kill predators for several decades, the official program of Predator and Rodent Control (PARC) was not established until 1915. With the prodding of a strong agricultural lobby, the budget for the program grew. By the 1920s, the Biological Survey was killing thirty-five thousand coyotes a year (Dunlap 1988: 51). The arguments used to support the eradication were based on scientific ideas similar to the ones opponents to the control program were using. Supporters echoed Roosevelt's 1915 hunting argument, claiming that man had now replaced predatory animals as the primary carnivores of the land. Less predators meant less competition, not only for livestock but for wild game as well. The "varmints" being killed off no longer served any purpose in the "balance of nature," and because they could only do harm to human interests, they were viewed as having negative value. Opponents to the eradication program were mostly mammalogists and zoologists who viewed the slaughter with a carefully couched moral outrage. The arguments most commonly presented used the "balance of nature" argument more ecologically and less managerially than the Survey administrators. Killing the predators, the scientists claimed, would permanently upset the checks and balances that nature had provided. Human hunting alone would not keep game at healthy population levels. Unfortunately, in the 1920s there were few scientific studies to back up this position. The eradication program was driven chiefly by the commercial interests of the stockmen and the negative value that they placed upon large predators.

To combat the negativism and the obvious support that the ranchers had within the federal government, scientists countered with familiar arguments for the scientific value of living organisms. One notable example was an article by Lee R. Dice, published in the *Journal of Mammalogy* in 1925, "The Scientific Value of Predatory Mammals." Dice declared that "every kind of mammal, as well as every other kind of organic being, is of great scientific significance, and the world can ill-afford to permit the extermination of any species or subspecies" (Dice 1925: 25). He argued that the loss of a species would mean the loss of anatomical, evolutionary, and ecological understanding: "There is no substitute in scientific work for the living animal and the fresh specimen, and every species exterminated marks a decided loss for the scientific world" (Dice 1925: 25). He especially emphasized the predator's ecological importance, in that removing the predator would disrupt relationships between the other living organisms of a particular region: "The lives of all species of animals living in one locality are closely interrelated; especially close are the relations between the carnivores and the forms on which they prey. All of these associated forms, predatory and non-predatory alike, have evolved under mutual adjustment, and all of these associates must be considered together in any attempt to explain evolution or distribution. With the predatory mammals eliminated it will become more difficult to explain the origin of many adaptive structures and habits in the remaining species" (Dice 1925: 27).

From the other side of the argument, in the same issue of the *Journal of Mammalogy,* E. A. Goldman of the Biological Survey offered an opposing view in an article entitled "The Predatory Mammal Problem and the Balance of Nature." Goldman asserted that after centuries of "occupation of the continent by Europeans bearing firearms, clearing the forests, and settling permanently throughout its [the continent's] extent, the balance of nature has been violently overturned, never to be reestablished" (Goldman 1925: 29). Goldman's conclusion was that the natural checks and balances in the relationships of different species had been permanently altered, and thus "practical considerations demand that [man] assume effective control of wildlife everywhere" (Goldman 1925: 31). Goldman scattered various statistics throughout the article, from the monetary value of livestock lost to predators, to the number of people who contracted rabies after being bitten by coyotes. He dismissed the assertion that large predators helped to keep rodent populations in check, although the only evidence he gave was his personal conversation with a "predatory-mammal hunter who has killed dozens of mountain lions," who reported that the stomachs of his kills contained nothing but the remains of deer. This was hardly the rigorous scientific study that would lay any debates to rest. Goldman also used this anecdote as evidence that predatory mammals are detrimental to game populations and ought to be controlled to allow more economically and recreationally valuable species to flourish. "Careful distinction must be made between species which are useful and harmless and their perpetuation to be fostered, and species which are so destructive that their effective control becomes an economic necessity, as well as in the interest of other valuable forms of wild life" (Goldman 1925: 32).

Goldman concluded with this statement: "Large predatory mammals, destructive to livestock and to game, no longer have a place in our advancing civilization" (Goldman 1925: 33). Although he described himself as a nature lover, "loath to contemplate the destruction of any species," he believed that the "practical" considerations outweighed any arguments against predatory mammal control. Goldman also asserted that fighting for protection of predators in areas "occupied by the homes of civilized man and his domestic animals" would turn popular opinion against other more important wildlife conservation efforts in the country. Goldman's position, representing that of the Biological Survey, emphasized not only the negative value of the predators but also the primacy of economic value of domestic animals and game. The predators' time had come and gone.

Invoking Values to Offset Negativism

It should be noted that not all scientists stayed strictly to the scientific value argument to counter the "practicality" of the Biological Survey's position. One prominent biologist, Dr. Charles C. Adams, in an address to the American Society of Mammalogists in 1924, proposed a comprehensive plan for conserving predators on the public lands of the United States. His paper (based on his address), "The Conservation of Predatory Mammals," listed the numerous values that would be lost if the present extermination practices were allowed to continue. "This discussion," Adams wrote, "is built on the assumption that these animals are worth preservation, *somewhere*," and he proceeded to "summarize the main reasons for utilizing predatory mammals for human welfare" (Adams 1925: 86).

At the top of his list was "scientific value," which Adams described not only as the importance of predators for the investigation of anatomical, ecological, geographic, taxonomic, and evolutionary questions, but also in terms of their importance in the "economy of nature," in particular as controls for the "herbivorous animals." Adams then moved on to "educational and social values," including the "recreational value" of predators as evidenced by the public interest in predator exhibits in "zoological gardens," as well as "moving pictures" (films) about "bears, wolves, foxes, and similar animals." Adams also claimed that the results of scientific studies, providing such concepts as parasite, scavenger, and predator, "have considerable educational value and are not wholly useless in their application to social relations." Nature, Adams claimed, had much to teach us about ourselves (Adams 1925: 86).

But the most interesting section of Adams's list of values was entitled "economic values," for here Adams directly challenged the major rationale for the Biological Survey's eradication program. Livestock was not the only economic interest impacted by predator control. Adams first returned to the issue of rodent overpopulation, a benefit of protecting predators that the Survey flatly denied. But more significantly, Adams decided to emphasize "the economic possibilities which have not yet been discovered. . . . With new conditions continually arising, old values are constantly changing. For years, the herbivorous muskrat was not considered of much value, but now all is changed, and we can never tell when there will be such changes. . . . Recently insulin, the priceless boon to the diabetic, has been found in the liver of the predacious shark. Who knows what yet remains to be found in the numerous predacious species which will be of priceless value to man? . . . If they are not preserved, and are allowed to become extinct, there is no hope whatever of recovering them" (Adams 1925: 87).

Adams was taking the economic argument and turning it back on the Survey. Predators, he claimed, were far from useless in the civilized world; the facts demanded that responsible humans work carefully to preserve them. And as Adams's colleague Dice had implied, such values applied to all species. Adams's economic values argument presaged the economic arguments for preserving biological diversity that would become popular in the late 1970s and early 1980s. In the face of such potential value, the "practical" solution (in Goldman's words) seemed a matter of prudence rather than politics.

But even the prospect of future economic value was ultimately not

convincing to those in charge, and the Survey claimed that the predator species targeted in their campaign would survive in other, less human, parts of their range. The rodent control issue remained controversial for many years, and the Survey still claimed that predator
control meant a healthy increase in game populations. With a lack of
scientific studies to prove or disprove any of their assertions, Survey
administrators did not change their minds until the famous Kaibab
deer saga.

The Kaibab Controversy

Described by one historian as the "classic conservation horror story,"
the fate of the Kaibab deer in the mid-1920s became a wildlife management fable that was to be repeated in sport periodicals and textbooks for many years (Dunlap 1988: 65). Although the causes of the
deer population crash would continue to be debated for many years,
the event clearly had an important influence on conservation policy.

The Kaibab Plateau, located on the northern rim of the Grand
Canyon, had been designated a game reserve in 1906 and managed by
the Forest Service since 1908. In attempts to raise the number of deer
for the benefit of hunters, the Forest Service asked the Biological Survey to employ their expertise in predator control. The Survey com-
plied, in subsequent years working to remove the wolves, mountain
lions, bobcats, and coyotes of the Kaibab. By the mid-1920s, it was evident that the protection policies to raise deer numbers had worked
too well. The rangeland on the Kaibab had been browsed to bare
ground and twig. With government scientists reporting the extensive
damage and recommending immediate action, the Forest Service decided that the only way to solve the problem was to open the refuge to
hunting. But the sporting community had convinced itself over the
years that deer populations were low and that preserves and predator
control were necessary to maintain higher numbers of game. In addition, Steven Mather, director of the National Park Service, used his position and the popular sentiment for protecting deer to oppose any
killing on the refuge. The Forest Service hunts were subsequently de-
layed.

But the severe winter of 1924–1925 forced the issue. Large numbers
of deer died, weakened by the lack of forage and their atrophied phys-

ical conditions. The sporting community voiced its protest and the state of Arizona (wildlife on federal lands was still considered under state jurisdiction) continued to block the Forest Service's attempts at organizing hunts. The sportsmen accused the government of pandering to "game hogs" who would happily take advantage of relaxed restrictions without any regard for the future. In 1928, the Supreme Court decided that the Forest Service could act to manage the deer on federal land without Arizona's consent, although this decision did little to change the perspective of opponents to deer hunting on the Kaibab. In legislative history, it was an important precedent, opening the way for federal control of wildlife on federal lands regardless of state laws (Dunlap 1988: 67–68).

At the time, the government would not publicly speculate on the connection between its predator policy and the ill health of the deer herd. In fact, the lack of predators to control the deer was likely only one of many factors precipitating the Kaibab fiasco (Caughley 1970). But the repeated lesson for the future generation of wildlife managers was clear. As Durward Allen wrote in 1954, "Many people, no doubt, think that freedom from predation would be the life beautiful for nearly any of our game" (D. Allen 1954: 234). He then recounted the Kaibab story to demonstrate that less predators had not resulted in more game. But at the time of the Kaibab hunts, the government would not admit that their predator policy had possibly been a mistake. Most still believed that natural predators would have decimated the deer herd and that human management was the only solution in the civilized world. Slowly, the predators' role in the economy of nature would become more clear as wildlife management became more of a science rather than an exercise in trial and error. It was, in fact, the rise of more methodical and quantitative studies of game in the 1930s that made wildlife management a respectable science, and "facts" began to displace management myths as a new discipline was established.

Elton, Leopold, and the Profession of Wildlife Management

One book in particular provided both inspiration and direction for the first generation of wildlife biologists in their endeavors to discover general laws of managing wild populations. In 1927, Charles Elton, a

British zoologist at Cambridge University, published *Animal Ecology,* a book about "the sociology and economics of animals" (Worster 1994: 295). Elton offered the American game managers a theoretical framework for producing detailed studies on specific animal populations. Whereas before, scientists had depended upon the idea of the "balance of nature" in their defense of predators and other "useless" animals (including birds) without any real evidence to support their claims, Elton's theories of "food chains," "trophic levels," and "niches" provided the foundation that encouraged quantitative studies of animal life histories. Elton believed that a careful analysis of each natural community would reveal a basic economy that existed among the living organisms. The different reproductive characteristics of the animals, the climate, the terrain, and the attendant plant life varied, but the general laws that linked together "producers" and "consumers" (as Elton called plants and animals) within the food cycle stayed the same.

Armed with these new ecological theories, scientists were motivated to go into the field and collect quantitative data that would support Elton's ideas. It was a new way of perceiving wild nature. Perhaps even more important, it implied that animal populations adhered to general mechanisms that could be managed for the benefit of humans. The craving for efficiency that had characterized the conservation of forests, water, and mineral resources during Roosevelt's administration was extending into the ideas of wildlife management. With the greater understanding of biological workings came the impulse to exert greater control.

The conservation impulse that Elton initiated was undoubtedly articulated most completely in Aldo Leopold's well-known *Game Management* (1933b). Published during Leopold's inaugural year at the University of Wisconsin as the first professor of game management in the United States, this book was considered the primary textbook for academic wildlife biology training well into the 1960s. Trained as a forester, Leopold spent the first years of his career with the Forest Service in the Southwest, where he built a reputation as an expert on wildlife and game issues. In 1925, Leopold transferred to the Forest Products Laboratory in Wisconsin, but after three years he resigned to work as a consultant for the Sporting Arms and Ammunition Manufacturers Institute. With the institute's financial support, Leopold began to dedicate all of his time to studying the game populations of the region (Trefethen 1975: 214).

One of Leopold's studies focused on the bobwhite populations of southern Wisconsin. Leopold was familiar and impressed with the work of Herbert Stoddard, a Survey biologist who had performed an intensive, quantitative analysis of the bobwhite population of Georgia starting in 1923. The resulting book, *The Bobwhite Quail: Its Habits, Preservation, and Increase* (1931), is often cited as the first practical example of scientific application in game management. Leopold, who was teaching part-time at Wisconsin in 1929, had his student Paul Errington follow Stoddard's methodology for the bobwhite of the north: carefully counting population numbers, nests, and clutch sizes; taking notes on food availability and weather patterns; and observing mortality rates and causes of death. Leopold used Stoddard's and Errington's work to help show how game population numbers were controlled in nature by identifiable factors and therefore could be easily manipulated by humans for beneficial results (Dunlap 1988: 72–73).

Although Leopold has come to be considered by many to be a prophet for modern environmentalism, a founding father of environmental ethics, and a champion of wilderness and holistic ecological values, his work up to the publication of *Game Management* in 1933 was solidly rooted in the Progressive utilitarian tradition of the early twentieth century. It is interesting that in this same year Leopold also wrote and published "The Conservation Ethic," an article that some feel signaled the beginning of a new way of thinking for Leopold, marking the start of his biocentric views of the natural world (Oelschlaeger 1991: 218). While we may catch glimpses of this new ethical sensibility in *Game Management,* the book-length work was chiefly dominated by an agricultural perspective replete with utilitarian and traditional recreational values. As Leopold clarified, "Game management is the art of making land produce sustained annual crops of wild game for recreational use" (Leopold 1933b: 3). This perspective, according to Leopold, represented a whole new "concept" of sport. Instead of the prevalent idea that "all hunting is the division of nature's bounty," Leopold suggested "that hunting is harvesting of a man-made crop, which would soon cease to exist if someone somewhere had not,

intentionally or unintentionally, come to nature's aid in its protection"
(Leopold 1933b: 210).

The "aid" described by Leopold took the form of various methods
designed to encourage the propagation of desirable species. The theory behind these methods was that there was often a limiting factor
that suppressed populations. By identifying this factor and working to
relieve its influence, managers could increase populations and provide
a crop for hunters to harvest. As Leopold wrote, "Game management
consists largely of 'spotting' the limiting factor and controlling it"
(Leopold 1933b: 39). By organizing his text according to such general
properties and factors, Leopold hoped to move beyond the more specific publications that focused on singular species or land types. His
goal was "to portray the mechanism which produces *all* species on *all*
lands, rather than to prescribe procedures for producing particular
species or managing particular lands" (Leopold 1933b: viii).

It is not clear what Leopold meant by "all species"; he probably
meant all *game* species. Certainly the examples he employs throughout
Game Management strictly involve animals desirable for hunting.
However, near the end of the book, Leopold did include a brief section, entitled "Management of Other Wild Life," in which he declared:
"The objective of the game management program is to retain for the
average citizen an opportunity to hunt. . . . It implies a kind and quality of wild game living in such surroundings and available under such
conditions to make hunting a stimulus to the esthetic development,
physical welfare, and mental balance of the hunter. The objective of a
conservation program for non-game wild life should be . . . to retain
for the average citizen the opportunity to see, admire and enjoy, and
the challenge to understand, the varied forms of birds and mammals
indigenous to his state. It implies not only that these forms be kept in
existence, *but that the greatest possible variety of them exist in each
community*" (Leopold 1933b: 403; emphasis Leopold's).

This could be considered an early expression of concern for diversity, although Leopold was selective in choosing only "birds and mammals" as valuable nongame species. Still, Leopold's emphasis on aesthetic development for both hunters and nonhunters reveals the broad
sensibility of the value of wildlife upon which Leopold would later expand. He saw this shared aesthetic (as Roosevelt had observed nearly

twenty years earlier) as evidence of "a fundamental unity of purpose and method between bird-lovers and sportsmen" (Leopold 1933b: 405), and he believed that such unity, if recognized, might serve to advance conservation efforts. Still, *Game Management* remains a book about manipulating nature for hunting, and in its narrow focus on game species, Leopold chiefly ignores both potential philosophical conflicts and the possibility of negative managerial impacts on non-game species. Leopold himself, in later works, contributed to the broadening perspective of wildlife that has more recently characterized the management profession he helped to establish.

Ding Darling and the First North American Wildlife Conference

Leopold's manifesto for wildlife managers could not have come at a more critical time. In the 1930s, the most popular representatives of endangered wildlife—migratory waterfowl—were suffering annual deficits in their population numbers, and the ratio of hunters to available game was becoming increasingly lopsided. In addition, the drought of the 1930s was aggravating the crisis. One source estimated that the U.S. duck population plummeted from a hundred million in 1930 to twenty million in 1934. Refuges were seen as the essential remedy, so a 1934 report from the President's Committee on Wildlife Restoration proposed an outlay of fifty to seventy-five million dollars for seventeen million acres of refuge lands. Needless to say, Depression-era budgets, although sympathetic to large government programs, were not ample enough to support such an ambitious purchase (Fox 1981: 191).

The waterfowl found their champion in Ding Darling, an avid duck hunter himself, who after years of jabbing criticism as an Iowa-based national cartoonist was pulled into bureaucratic service by an old friend, Secretary of Agriculture Henry Wallace. In 1933, Wallace convinced Darling to head and reorganize the Biological Survey. Darling, who had no experience as an administrator, agreed to take the job for six months, although he ended up staying for nearly two years. He proved to be effective in squeezing money out of Congress, as well as leading the fight for the Duck Stamp Act of 1934, which required every hunter to buy a postage-style stamp, the proceeds of which went into the fund for purchasing refuge land. Many conservation histories speak about him, in glowing terms, as a champion of sport and wildlife.

There were those, however, who criticized Darling for his seemingly single-minded focus on ducks and strong support of hunting. For example, some in the protectionist constituency viewed the Duck Stamp Act as a way to avoid setting lower bag limits and shorter seasons. The periodical *Nature Magazine* published several editorials and articles on this topic. In one piece, "The 'New Deal' for Waterfowl," the editors wrote that "the ostensible purpose of the waterfowl stamp is to furnish funds for the purchase of refuges. We believe that its real purpose is to stifle restrictions" (*Nature Magazine* 1934: 197). Others saw little change in the Biological Survey's PARC program and believed that Darling, who had once been an ally of wildlife defenders, had been caught in the machinery of the federal bureaucracy. Perhaps because of these public attacks from people and groups that he once considered partners, Darling resigned his post in late 1935, content to know that in the time of his government tenure he had presided over the acquisition of nearly one million acres of refuge land for his beloved ducks (Fox 1981: 194–196).

But Darling still had a significant role to play in wildlife circles. While at the Biological Survey, Darling had helped to coordinate the formation of the American Wildlife Institute (AWI), a new conservation group designed to represent the business interests that benefited from hunting. Darling's ostensible interest was to use the influence and money of the AWI to organize a huge wildlife conference that would look at all facets of the "crisis" in wildlife resources. With administrative help from the government and public support from Franklin Roosevelt, the First North American Wildlife Conference took place in February 1936. It has since been considered a watershed event for stimulating conservation action. As a result of this conference, the General Wild Life Federation, an umbrella organization for the thousands of wildlife groups that endorsed the conference, was formed in 1938 and Darling was elected as its first president. This group, which helped to organize annual wildlife conferences in subsequent years, would later change its name to the National Wildlife Federation.

Still, the ties of the AWI to the gun lobby and business interests bothered many nonsportsmen, and indeed the conference almost exclusively focused on huntable birds and mammals. Although official speakers celebrated the many values of conserving wildlife—not only economic, but also recreational, educational, social, and even spiritual—such values were always viewed in terms of the pursuit of game (*Proceedings* 1938: 31, 36, 38). But there were a few individuals who represented the nature lovers' constituency at the 1936 conference. One was John H. Baker, the president of the Audubon Association, who spoke out against predator extermination, declaring that "such control results in the destruction of many forms of wildlife in the belief that this is desirable in order to produce . . . suitable targets or salable birds or animals. . . . Other people in the community may well attach as much or more value to the birds and animals being controlled" (*Proceedings* 1938: 138). In addition, there was one special session, the "Problem of Vanishing Species," which included presentations on the trumpeter swan, the Sierra bighorn sheep, the sandhill crane, the grizzly, and the pronghorn antelope. Even William Temple Hornaday, the grand old man of the antihunting forces was asked to speak, and he responded with his usual flair, denouncing "the devastating power and deadliness of the . . . grand armies of well-armed killers who each year go out against the cowering remnants of small game, and kill all of it they can find" (*Proceedings* 1938: 663–664).

Thus, the conference did indeed represent an attempt at bringing together people with an interest in conserving wildlife. Darling saw it as a first step toward trying to unite forces under a common goal. "Wildlife interests," he said, "remind me of an unorganized army, beaten in every battle, zealous and brave, but unable to combat the trained legions who are organized to get what they want" (quoted in Fox 1981: 196). Although the hunters and the nature lovers apparently would never see eye to eye, the conference brought wildlife issues to the attention of the nation's public.

An Appreciation for Aesthetics

To talk of protecting the aesthetic value of the plant and animal world, as Leopold and Roosevelt had attempted to do, served to link diverse interests. While no one could offer a strict definition of *aesthetic* that would cover all usages, many writers employed it to describe a kind of intangible value other than economic or scientific value, or to capture a characteristic of the human experience in nature that could not otherwise be expressed. Perhaps stimulated by the first Wildlife Conference, defenders of aesthetics began to speak out, and as a result, more species became objects of conservationists' concern.

The year after the Wildlife Conference, in a 1937 issue of *Nature Magazine,* William Rush addressed the issue of wildlife's aesthetic value in an article entitled "What Are Wildlife Values?" The piece is significant not only because it called attention to the value of nongame species, or because it emphasized the importance of aesthetic value in management decisions, but also because Rush was the director of wildlife management for the Biological Survey in Washington, Oregon, California, and Nevada. Rush was not a sentimental nature lover proclaiming the evils of hunting or a mammalogist denouncing predator control. He was a professional offering his thoughts on why a range of wildlife should be protected for the benefit of all American citizens. He even offered a definition for aesthetics—"the science of the beautiful in Nature and art"—and he made clear that its significance should not be ignored just because it was easier to calculate the "dollars and cents value" of wildlife.

Rush began by commenting that "a good deal is said about the values of wildlife . . . but usually little or no value is assigned to the nongame species" (Rush 1937: 40). He observed that values were most commonly divided into four categories: economic, recreational, educational, and aesthetic. But, recognizing the popular emphasis on quantitatively measuring wildlife value by using hunting expenditures as a proxy, or the "paltry value of the meat of these animals as a food resource," Rush asserted that "in our zeal to put a dollar mark on everything, we are losing sight of the real, more lasting values derived from the wild things of field, forest, and waters" (Rush 1937: 40). He concluded that the value priorities are completely turned on their head: "If we accept the definition of 'value' as 'the properties of a thing in virtue of which it is useful or estimable,' then the time-honored list of values of wildlife named on the old order of importance as economic, recreational, educational, and esthetic should be reversed" (Rush 1937: 40).

Rush went on to argue that "the appreciation of grace and beauty is latent in all of us, even in the most hardened hunters and fishermen," that aesthetics is the one value shared by all who wish to protect wildlife, and that therefore it deserves primacy on any list of values. After describing the benefits of the "eternal mystery of Nature" as the "inspiration to poets and artists," and as fostering "the appreciation of [the] graceful and beautiful," Rush reiterated that "none of these good things can be evaluated in money," although he stated clearly his opinion that most wild animals "are worth a hundred times more alive than they would be if dead." He concluded, "Life is much better on this earth because of the presence of these wild things that delight and please us by their mere existence. Surely the esthetic value of wildlife should be given first place in the scale" (Rush 1937: 41).

There is ample evidence that other writers in the late 1930s and early 1940s agreed with Rush and expanded on his ideas. V. H. Lehmann, writing for *Bird Lore,* contributed an article in 1938 entitled "Some Values of Natural Areas," in which he emphasized the aesthetic importance of species preservation. After summarizing the scientific values of undisturbed areas "as a vast outdoor laboratory," Lehmann continued: "But practical values from natural areas would be surpassed, I believe, by the esthetic gains. Conservationists look on natural areas as the only means of saving some species of animals and plants from otherwise certain extinction" (Lehman 1938: 312). In addition to advocating the protection of species that contribute to the beauty of the world, Lehmann was implying that the aesthetic value of wildlife demands that we prevent extermination of those living organisms not as dramatically "graceful" as those emphasized by Rush.

Ira Gabrielson, the successor to Darling as director of the Biological Survey, published *Wildlife Conservation* (1942), a book in which he included a chapter on managing for nongame species, emphasizing the importance of aesthetic value. He made special note of the lesser-appreciated "living forms," including "the leap of a salmon, the swift swoop of a hawk, the delicate tracery of a spider's web, the sheen of a butterfly's wing . . . and countless other characteristics [which] appeal to the esthetic in humanity" (Gabrielson 1942: 169). By listing such examples, Gabrielson implied that management strategies for wildlife must take into account other factors besides those that simply help to propagate game.

One would not expect to find aesthetic arguments for preserving nature in articles that were published during World War II. But Alan Devoe's piece "On Salvaging Nature," published in 1944 in the popular periodical the *American Mercury,* argued against the narrow focus on conserving natural resources only for their material usefulness in winning the war: "The conservation of petroleum is clearly and evidently linked to the war. But so, more subtly but none the less surely, is the conservation of trout, trees, meadows, mountains, and every whooping crane and roseate spoonbill." Devoe discussed two reasons to support his statement. The first reason was what he called "simply practical and statistical," interpreted in economic terms as future commodities. But the second reason, he claimed, was "a better one: one that has to do with quite another kind of values." Devoe went on to paint a picture of young American men returning home after the war only to find a devastated and neglected natural world. Such an impoverished outdoors is not what they will want or deserve. What they will need is "an outdoors at least as clean and flourishing as when they left. . . . They will want to get back to the American earth: to take the old fishing gear from the closet and dust it off and set out for some shaded, swirling pool where the trout are. They will want, most urgently, to take up the gun of peacetime instead of the gun of war. They will want, in their weariness and for the cure of wounds of the spirit, just to enter the healing place that is the forest, and stand in silence and aloneness there, and be renewed" (Devoe 1944: 369).

Thus, an argument for the salutary powers of nature's aesthetic value likely played well to a public craving an end to the war and a safe return of their families. It is also evident in Devoe's words that he considered the outdoors and its inhabitants to be a cultural symbol of America's freedom and beauty. In this way, aesthetic value was no longer limited to bird-watchers and humanitarians. It was within the purview of all who had a stake in protecting and conserving the natural world.

Finally, an article published in 1948 perhaps best represents the full articulation of aesthetic value as a reason for protecting species. The article was by Alexander Skutch, a prominent ornithologist and naturalist, and it was published in *Audubon Magazine*. In one passage, Skutch wrote poetically not just of the beauty of nature but also of how an aesthetic appreciation could open one's eyes to the variety of life that surrounds us. Exploring the question of what constituted the basis of the aesthetic value of wild places, Skutch wrote:

> It depended in the first place upon the multitude and diversity of the living creatures that surrounded them. The centuries-old tree with its

towering trunk and massive boughs was no more essential than the delicate herb with its frail ephemeral flowers and the green moss that car-
peted the rocks. The deer and the bear added much to the mysterious
delight of the woodland, but no more than the shy bird with its gay
plumage and ringing song and the saucy squirrel that scolded from the
bough overhead. The butterfly with its brightly painted satiny wings
contributed its mite to the total impression, but so to did the spider
with its tediously spun insubstantial web and the milliped that lurked
in the decaying fallen log. Thoughtful people discovered that much as
any one of these divers organisms offered study and contemplation, it
gained immeasurably in interest when considered in relation to all the
other organisms, both animal and vegetable that surrounded and inter-
acted with it. (Skutch 1948: 358)

For Skutch the diversity of the biological world—small or large,
hidden or conspicuous, timid or bold—provided the magic that de-
lighted the human eye and mind. Skutch identified this powerful aesthetic value as a central reason for protecting animals and plants in nature. As he concluded, "If then we are to conserve our resources and
thus prolong our lives upon this earth, we must base our conservation
philosophy upon our planet's twofold fitness for human life. We must
maintain its ability to supply our complex physical needs and, at the
same time, keep earth's beauty for growth of our esthetic capabilities"
(Skutch 1948: 359).

Aesthetic value, at least in some circles, had found parity with the
utilitarian value of nature. As Skutch and Devoe both implied, the
nourishing element of nature's beauty should neither be overlooked
nor underestimated. Aesthetics, for those willing to open their eyes,
was considered a valuable benefit that all citizens could enjoy.

Cultural Value

A second value that became more prominent in this period after the
Wildlife Conference was one that connected wildlife to the cultural
character of America and its citizens. Devoe had hinted at this value
when he called for the preservation of nature so that the returning soldiers could enjoy the pristine outdoors, an experience that would re-
connect them with their American roots. But this cultural value of
wildlife was represented even more pointedly by Edward H. Graham,

in a 1947 article in *Audubon Magazine* entitled "Wildlife Is Part of Our Heritage." Graham was chief of the Biology Division of the Soil Conservation Service, and this position—examining the biological aspects and effects of soil conservation for an agency that worked closely with the American farm community—likely inspired his thoughts on the importance of animals throughout history to the development of human culture. As he wrote, "We must admit that it is virtually impossible to imagine civilization without domesticated animals, the influence of wild animals in literature, art, and architecture, and their part in religious and allegorical symbolism" (Graham 1949: 105). This contribution for Graham certainly was not just limited to game or farm animals. It was the experience of generations that grew up in a rich, complete, and abundant natural world that provided the cultural foundation on which Americans—and in fact, all peoples—based their identities. For this reason, Graham said, we have an internal, emotional reaction when faced with the prospect of a species being exterminated: "It is no wonder, then, that there is within us something that cries out when wildlife is endangered, as when a species is threatened with extinction, or when marshlands, forests, and other homes for wild birds or mammals are materially changed or destroyed." This was not just an issue for special interest groups: "It is not the voice of the hunter, who loses game to shoot, nor the naturalist, who loses a species to study, that alone challenges change." Graham saw a value "far stronger than the recreational and economic considerations so often stressed today" (Graham 1949: 105): the deep internal reactions of the people. Connecting conservation values to the American character made wildlife conservation a significant issue for everyone. As Graham concluded, "Our whole history as a nation, people and race, sees in the loss of wildlife an injury to our cultural heritage we cannot easily tolerate" (Graham 1949: 105).

Perhaps no creature in the wild better represents America to the people than the bald eagle. As the national symbol, the eagle's likeness is pervasive throughout our society: adorning buildings, gracing currency, symbolically positioned in historically significant artwork, prominently displayed on official government seals. Ironically, however, the eagle had been a target in the general extermination campaign against predators because of its appetite for game birds and smaller members of a rancher's domestic stock. In 1940, Congress enacted the Bald Eagle Protection Act, which made it a criminal offense "to take or possess any bald eagle, or any part, egg, or nest thereof" (Bean and Rowland 1997: 90). The law took its lead from the 1918 Migratory Bird Treaty Act in its prescribed penalties and limited exceptions. A law protecting the national symbol would not have met much opposition in Washington, DC, especially with the war looming. But there was an important issue in this case about federal jurisdiction over wildlife and what powers Congress could claim to protect the eagle. Previous wildlife laws had used either the power to regulate commerce (as with the Lacey Act) or the power to engage in treaties (as with the Migratory Bird Treaty Act). Neither of these powers seemed to apply to the case of the bald eagle. Instead, Congress declared that the federal government had the authority to preserve national symbolism (Lund 1980: 50). In these terms, it was likely difficult to oppose the act as unconstitutional. But by circumventing the traditional bases for federal wildlife protection, Congress made clear that there were other values besides those traditionally cited for protecting wildlife. It was the first time that the stated primary reason for protecting a species was a value other than economic, recreational, or scientific.

Ecological Awareness: Interconnections and Interdependence

At the same time that aesthetic and cultural values were being used as motives capable of uniting wildlife conservationists, another more scientifically based perspective was gaining in popularity, one that would emphasize the importance of a full complement of wildlife for the maintenance of a healthy natural world. Many writers began to focus on the ecological value of wildlife, a value based on the traditional idea of the "balance of nature." Charles Elton had provided a major step forward in *Animal Ecology* (1927), by proposing theories of how animal populations were mediated by both their biological capabilities and their environmental conditions. The next stage was to understand nature as a system of component parts, each playing an important role in the greater network. In 1935, an Oxford botanist named Arthur Tansley published an influential essay in which he proposed the idea of the "ecosystem," which was an attempt to capture all of the physical connections between living and nonliving constituencies in a particular area. Tansley, however, was largely dismissive of such holistic ideas as "communities" or representations of nature that emphasized an emergent whole over the study of the component parts. His interests lay in more reductive analysis, whereby the parts are distinguished and the links between them are scientifically explained. Tansley was partly responding to the "organicism" trend of the early twentieth century, a perspective that he believed was keeping ecology from being taken seriously as a scientific discipline. The "ecosystem" was thus partly intended as a way to reduce the relationships of plants and animals to simple expressions of energy exchange, a much more quantitative, mathematical, and therefore "scientific" representation of the natural world (Worster 1994: 301–302).

But the reductionist trend that Tansley's ecosystem concept represented did not completely pass into the popular view of nature. Instead, the combination of wildlife biologists performing more in-depth studies of specific populations and the scientific idea that each species had a role to play in the greater energy transfer mechanism of the ecosystem encouraged a new appreciation for a broader complement of plants and animals than simply those considered "useful" or "beautiful." In fact, the whole idea of categorizing flora and fauna as "good" and "bad" started to be viewed as outdated and misleading. In a telling example, E. L. Scovell published a short essay in *Recreation* in 1938, entitled "Overlook No Living Thing." Not only did Scovell emphasize the importance of not judging animals and plants in the face of our lack of ecological knowledge, but he also more pointedly tied human well-being to the fate of other living organisms: "Man cannot escape his dependence on all forms of life. Yet here is a point that is rarely if ever stressed in programs on conservation. All too often the stress is placed on the good and the bad plants, animals, birds and insects; on the good and the bad things a person can or should not do. Who knows what is good or bad, friend or foe? Under certain conditions a plant, animal, bird, insect, fish, or reptile may be an enemy. Under other conditions it may be a real friend. . . . We cannot overlook any species of living thing from grass to trees, from the smallest insect to the largest bird, from the smallest mouse to the largest animal, from the tiny minnow to the whale, from the lightest rainfall to the largest body of water" (Scovell 1938: 295–296). Scovell here expressed a concern for the complete range of species similar in character to Skutch's concerns and also to later concerns for biological diversity, and this viewpoint, although more romantic than some scientists might prefer, was definitely informed by an ecological view of the natural world.

The same idea, however, was being represented in more scientifically based publications as well as in the professional circle of wildlife management. One example is Gabrielson's 1942 book *Wildlife Conservation.* As director of the Biological Survey, Gabrielson serves as an excellent representative of the approaches that were popularly accepted at the time. His book was very much in the spirit of Leopold's *Game Management,* although it was broader in scope, giving the basics of general conservation of soils, waters, and forests in relation to the conservation of wildlife. In a section entitled "Some Basic Facts in Wildlife Conservation," Gabrielson repeated the same ideas that Scovell had expressed earlier: "It is to be emphasized . . . that from the purely biological point of view there are no beneficial and no harmful plants or animals. All are cogs long fitted in the great biological mechanism, one having its function as well as another. The rattlesnake has its place as much as has the oriole; the worm does its turn in the drama of life as worthily as does the deer" (Gabrielson 1942: 106–107).

This statement of ecological value did not necessarily mean that humans should stop removing certain members of the animal world when they conflicted with human interests, as the Survey's embattled predator policy still dictated. But Gabrielson made clear that this was "an artificial classification of wildlife on the basis of self-interest" (Gabrielson 1942: 109). Without negative values impressed upon the natural world, all animal and plant members have value in the roles that they play for the greater whole: "On any other basis, all living forms, even weeds and the worst animal pests, are useful elements in the natural community and in the natural processes of building and maintaining fertile soils, conserving water, and providing an available supply of plant and animal food for all life" (Gabrielson 1942: 167).

An ecological way of thinking about wildlife was now rooted in the professional language of discussing reasons for conservation. It was the anointed successor to the "balance of nature" perspective of earlier years. By the time Leopold's *Sand County Almanac* was published in 1949, ecology could be said to hold an influential position in the American vision of the natural world. This new scientific paradigm, as Leopold would emphasize, was far more valuable to us than the less rigorous observations of our ancestors. "Ecological science has wrought a change in the mental eye. It has disclosed origins and functions for what to [Daniel] Boone were only facts. It has disclosed mechanisms for what to Boone were only attributes. We have no yardstick to measure this change, but we may safely say that, as compared with the competent ecologist of the present day, Boone saw only the surface of things" (Leopold 1949: 174).

As a result of the new view, Leopold observed, we had become more aware of the connections between the human world and the natural world. Leopold singled out this awareness as an important reason for conserving wild nature: "There is value in any experience that reminds us of our depending on the soil-plant-man food chain, and of the fundamental organization of the biota" (Leopold 1949: 178). In short, the ecological value of animals and plants demanded that we consider all living things in conservation, for they not only were dependent upon one another in the complex workings of nature but they also provided the foundation for the natural systems that supported human beings.

This ecological perspective carried over into the 1950s, and evidence suggests that it was starting to permeate traditional wildlife management discussions. For example, a book by John Black published in 1954 illustrates this broadening approach to species protection. The title—*Biological Conservation, with Particular Emphasis on Wildlife*—is itself revealing, as it implies a more inclusive approach to the conservation of the living world. As Black noted in the preface, "Heretofore in too many studies and books the term *wildlife* has had a restricted meaning limited to the birds and mammals of interest to the sportsman. In the present book *wildlife* is interpreted to include not only the game birds and mammals, but the nongame forms as well" (Black 1954: vii). In addition, Black did not limit himself to nongame birds and mammals, but as he made clear in chapter 1, "The Need for Conservation," all living organisms have a valuable role to play in the lives of future human generations: "We cannot expect to see millions of buffalo again. We can, however, expect that posterity shall find an abundance of birds, mammals, reptiles, amphibians, fishes, and invertebrates, each in its proper place, each contributing its little bit to the balance of nature, the maintenance of a healthy soil, and a healthy, happy people" (Black 1954: 6).

Of course, traditional game management concerns were still firmly established as the primary goals of wildlife conservation. But it was apparent that the wider range of values meant that a broader constituency now held interests in the preservation of species. In *Our Wildlife Legacy* (1954), with a short-paragraph writing style reminiscent of Leopold, Durward Allen focused for the most part on the concerns of sportsmen and combined scientific knowledge with down-home opinions about the state of conservation. While he was concerned primarily with the management of game, Allen (a highly respected professor from Purdue University) emphasized the importance of all wildlife as a national treasure that all citizens should appreciate. Although "game and fish are featured in these discussions," Allen wrote, "this book is intended to serve all groups who have an interest in developing strong state and national wildlife programs. In such a connection, civic groups, women's clubs, garden clubs, and many others are seeing their national function and their social responsibility" (D. Allen 1954: v–vi).

In this vein, ecology and nature study were enjoying a remarkable surge in popularity. Reminiscent of the turn-of-the-century era, when youths were encouraged to engage nature through specimen and egg collections, it seems that postwar suburban communities were seeking a reconnection to the natural world. The National Audubon Society, for example, ran educational programs at their nature centers, each of which, they claimed, served more than ten thousand schoolchildren and boy scouts each year. So popular were the programs in California, reported the National Audubon Society president in 1954, that Los Angeles schools had requested local centers to make room for fifty thousand children per year. The Audubon Society's primary message was clearly ecological and inclusive: "The Audubon Society disagrees with the idea still rampant that wildlife and plants can be classified as beneficial or harmful. . . . All wild plants and animals in their native environments have roles to play in the balance of nature" (Baker 1954: 620).

Science teachers also went to summer camps to learn how to incorporate ecological principles into their classes. Science institutes sponsored by the National Science Foundation in the 1950s trained biology teachers in the latest concepts about ecology and evolution. These sessions apparently also coincided with a change in some of the texts used. Carol Beall Stone, an educator who examined trends in biology textbooks in the twentieth century, noted the rising influence of ecology: "The most striking change to occur was the philosophical change reflected in biology textbooks—humans changed from 'masters of the earth' to 'man in the web of life'" (C. Stone 1984: 243). Surely this language both reflected and encouraged a more holistic, ecological view of the natural world and its inhabitants.

Thus ecology was not only firmly rooted in conservation thought; it also had found a niche in more popular interest in the natural world. This foundation contributed to later expressions of concern for the environment, but for the decade or so following World War II it largely lent itself to efforts to protect the inhabitants of the natural world that were being threatened by human activity.

Endangerment and Rarity

This particular impulse is best illustrated by the rise in concern for endangered species in the 1950s. Not all animals enjoyed the cultural status of the bald eagle, but it was apparent that in the years following the war, people were becoming more interested in species whose days seemed numbered. Hornaday and a few others had sounded the alarm in the first decades of the century, and at the 1936 Wildlife Conference the panel entitled the Problem of Vanishing Species had been a sign that attention to endangered animals that were not commercially or recreationally valuable was slowly gaining ground. With the new recognition of aesthetic, cultural, and ecological value, all species in the natural world now had a chance to be considered worthy of protection.

The *rarity* of some species was quite plainly what placed them in the spotlight. In a booming postwar economy in which population growth and commercial development were expanding at unprecedented rates, American citizens began to sense that the human power to alter the natural world was threatening whole segments of nature. The general awareness of the ecological connection between animal and habitat made it painfully clear that hunting was no longer the primary culprit in species decline. As cities and suburbs expanded, encroaching more and more upon natural areas, people sensed that there was a loss of contact with the wild America that only existed in collective cultural memories and in history books. This made them feel more drawn to stories about animals struggling for survival.

One species in particular that captured the attention of the American and Canadian public was the whooping crane. A rare bird even in colonial years, the remnant populations of whooping cranes were sent to the brink by hunters, egg collectors, agricultural developers, and builders who drained key breeding areas (Matthiessen 1987: 254). In *Our Vanishing Wild Life,* Hornaday had written that "this splendid bird will almost certainly be the next North American species to be totally exterminated" (Hornaday 1913a: x). By 1926, it was estimated that less than a dozen pairs were left. In 1937, the Department of the Interior purchased forty-seven thousand acres on the Texas coast as a migratory waterfowl refuge (the Aransas National Wildlife Refuge), largely because it was known that a small flock of whoopers used the area as a wintering ground. But by the 1940s, it was apparent that more needed to be done. Nothing was known about the northern breeding grounds, nor had anyone mapped out the migratory route that the cranes traveled each spring and fall. In an effort to gain what might be valuable information about the migratory habits of the species, the National Audubon Society and the Fish and Wildlife Service started the Cooperative Whooping Crane Project. Citizens in the United States and Canada were enlisted as "crane spotters" to try to identify migration routes and perhaps to help find the breeding grounds. The crane's fortunes were reported on the national news; *Life* magazine published a feature article; an oil company was persuaded to stop its drilling activities near the Aransas refuge (Matthiessen 1987: 256). The well-publicized project benefited from the drama and the aura of mystery that surrounded the rare birds. Where did the whooping crane disappear each summer? Would they return to their refuge in the winter, or meet some unknown fate? For ten years, the Crane Project collected reports, explored the northern reaches of Canada, and finally, in 1954, the whooper's summer breeding grounds were discovered in the already protected Wood Buffalo Park in northern Alberta. As Robert Allen, a director of the project, wrote, the crane's "dramatic struggle for survival has been so widely publicized that today no literate person in Canada or in the United States has failed to hear about it" (R. Allen 1960: 123).

As with the whooping crane, there was often a sad story associated with each species that was fighting for survival. These stories were ripe with drama and provided plenty of opportunity for the nostalgia that some segments of the population were seeking. The endangered species stories began to appear in articles published in environmental groups' magazines, often the same stories told over and over. The National Wildlife Federation had produced a list of about fifteen endangered species in the mid-1950s, and nature journalists seized upon the material. These species became early ambassadors of what would grow into a powerful, emotional, and controversial conservation issue. The cast of characters was always the same in these early articles: in addition to the famous story of the whooping crane, the plights of grizzly bears, trumpeter swans, California condors, key deer, graylings, ivory-billed woodpeckers, and sea otters now became familiar to those who read the environmental periodicals. It was almost prerequisite that these articles invoke the buffalo or the passenger pigeon to recall human transgressions of the past and that they express a nostalgia for when the country was more wild. For example, Will Barker's 1956 article "Our Vanishing Species," published in *American Forests,* provided synopses of species "whose existence is in jeopardy" and then concluded: "It may not be possible to save all the endangered species . . . but let us hope that the majority of them can be saved—living representatives like the once endangered buffalo of the wildlife America that Audubon, Douglas, and Muir knew and loved" (Barker 1956: 16). This was not simply an appeal to ecology, or science, or recreational sport. It was a direct appeal to a sense of heritage and history that was particularly inviting to Americans in the new postwar economy.

The Emergence of Environmentalism

In addition to bringing peace and prosperity, the 1950s also brought an overall change in the attitudes of Americans toward the environment. Historian Samuel P. Hays, in his book *Beauty, Health, and Permanence: Environmental Politics in the United States, 1955–1985* (1987), writes of a "transformation of values" that occurred in the postwar years: "This began with a rapid growth in outdoor recreation in the 1950s, extended into the wider field of the protection of natural environments, then became infused with attempts to cope with air and water pollution and still later with toxic chemical pollutants. Such activity was hardly extensive prior to World War II" (Hays 1987: 3). Hays asserts that these new environmental values "were clearly associated with rising standards of living and levels of education" (Hays 1987: 3). Far from being a radical fringe element in society (as environmentalists in the 1960s were often characterized), those whose concern for the environment was growing in the postwar years adhered to the traditional values of an upwardly mobile middle class. Environmentalism, in its infancy, was a "desire to improve personal, family, and community life" (Hays 1987: 5). As opposed to the conservation movement at the turn of the century, which was characterized by the drive for efficiency in the use of resources, the environmental movement "was more widespread and popular, involving public values that stressed the quality of human experience and hence of the human environment" (Hays 1987: 13). In fact, the actions of the more efficiency-minded managers now often clashed with those concerned with the overall quality of the natural world.

Hays says that the changing values reflected three aspects of environmentalism: the search for environmental amenities (given improved social conditions); the assurance of physical well-being; and an ecologically-based understanding of human dependence on nature. These three factors reflect the beauty, health, and permanence in Hays's title: "Whereas amenities involved an aesthetic response to the environment, and environmental health concerned a choice between cleaner and dirtier technologies within the built-up environment, ecological matters dealt with imbalances between developed and natural systems that had both current and long-term implications" (Hays 1987: 26–27).

Thus, Hays believes that the development of aesthetic and ecological values best characterized the newer environmental interest. In addition to widespread access to the beauty of nature through increases in disposable income and leisure time, improved technology began to clarify the impact that humans were having on the environment, at times revealing that some human illnesses could be traced to activities that had ignored the ecological principles that were now better understood.

The concerns that arose from the transformation of attitudes that Hays characterizes were encapsulated and given full voice in a book whose publication would mark for many the beginning of the modern environmental movement. Rachel Carson's *Silent Spring* (1962) painted a frightening picture for her readers: pesticides and other chemical compounds used by agriculture and industry were poisoning the land, the animals, and most shockingly, the humans who were supposed to be reaping the benefits of chemical use. All three of Hays's values— beauty, health, and permanence—were prominent in Carson's book. Her introductory chapter, "A Fable for Tomorrow," describes a make-believe town, once thriving with a diversity of plants, animals, and prosperous farms, that soon fell victim to an insidious "blight." Sickness was more common among both adults and children; farm animals were having difficulty reproducing; and a "strange stillness" hung over the once vibrant community. The animal life, most prominently the birds and insects, had vanished. "It was a spring without voices. On the mornings which once throbbed with the dawn chorus of robins, catbirds, doves, jays, wrens, and scores of other bird voices, there was now no sound. . . . The apple trees were coming into bloom but no bees droned among the blossoms, so there was no pollination and there would be no fruit" (Carson 1962: 2–3). The description of the lifeless town, complete with the fallout of a "white powder" on house roofs, lawns, fields, and streams called to mind the image of the aftermath of a nuclear bomb. But as Carson emphasizes, "No enemy nation had silenced the rebirth of new life in this stricken world. The people had done it to themselves" (Carson 1962: 3).

Carson goes on to spell out, in painstaking detail, the chemical make-up, characteristics, and apparent impacts of pesticides on both human and animal life. The book and Carson herself were immediately subjected to intense criticism. She was painted as a "hysterical" woman whose knowledge of science was suspect and whose emotional perspective obstructed any realistic and practical assessment of the situation. But as later studies would show, Carson was perhaps too conservative in many of her indictments. In addition, her careful research, which brought together the numerous but previously disparate reports of the environmental effects of chemicals, became recognized as incontrovertible evidence of how humans were endangering themselves and all life around them.

While the studies that linked human health issues to pesticides were more sparse and less conclusive then than they are now, the unintended impact of the various chemicals on nontarget wildlife had been documented in many cases. Silent Spring especially brought to light several ecological principles that previously had been unfamiliar to the

popular mind. One of these was bioaccumulation (or biomagnifica-
tion)—the way in which toxins enter the food chain at the lowest level
and eventually become more and more concentrated in each subsequent consumer, until lethal levels are attained. For example, Carson
described how in the 1950s western grebes in California had been
discovered with remarkably high concentrations of the insecticide
dichlorodiphenyldichloroethane (DDD) in their fatty tissues. The concentration of the actual treatment to control gnats in the lake where the
grebes lived was much smaller, by five orders of magnitude. But as Carson explained, the plankton organisms of the lake had absorbed the
DDD in higher concentrations, and thus the small fish that ate the
plankton had developed even higher concentrations. The deadly progression continued until the grebes, who are fish eaters, were exposed
to toxic levels far beyond what those dispensing the DDD could have
foreseen. Carson relates story after story of how chemicals applied to
water or soil enter into the food chain and persist for years after their
intended targets are long dead. Carson also uncovered evidence show-
ing that pesticides often became ineffective over time, as new generations of insects developed immunity to the treatments. Unfortunately
this display of evolutionary adaptability only convinced decision makers to apply stronger concentrations of the insecticides, often with
more severe effects for nontarget species.

More than any other book of the times, *Silent Spring* brought home
lessons of ecology to the average American citizen, and provided an effective critique of the new industrial management style of the natural
world. The controversy that arose around Carson's credibility and the
chemical industry's practices served mostly to bring the book's subject
to the attention of more people. While the dangers to human health
understandably elicited the greatest concern, the documented impacts
on fish, birds, and other wildlife helped to solidify the perception that
human civilization was now far more destructive than previously
thought. Nature and its inhabitants began to be viewed as a fragile
community under siege by an industrial human world. The evidence of
the interdependence of species highlighted the dangers that humans
were creating for themselves. If fish and birds were suffering from impacts that humans had not predicted, what hidden hazards lay in store
for human beings themselves? The overwhelming impact of *Silent
Spring* was to set humanity on notice: tend more carefully to the world

around you lest you find that you have destroyed the living fabric that provides all with life.

Federal Action for Wildlife: New Laws and the Leopold Reports

Around the time *Silent Spring* was published a flurry of new environmental laws and administrative policies were just emerging, and it seemed that Carson's book had at least partially provided the impetus to legislators and their constituents to demand more protection, not only for human health but also for the natural world. While laws like the Clean Air Act of 1963, the Solid Waste Disposal Act of 1965, and the Clean Water Restoration Act of 1966 all ostensibly were aimed at improving living conditions for humans, there were also obvious benefits to wildlife in the efforts to curtail pollution and clean up the environment. In fact, wildlife was often mentioned in the language of these laws as an intended beneficiary.

But numerous other laws were passed that had a more direct impact on the nation's flora and fauna. An example is the Wilderness Act, which had been introduced in Congress in varying forms since 1956 but had been defeated by opponents or by lack of interest for several years running. As one observer, writing in 1970, asserted: "The main conservation theme in 1960 was that America had to preserve the remaining wilderness regions and not open them to further exploitation. Actually, there was more involved in this stand than a mere nature lover's sentimentality. Man now had the means to bulldoze the entire continent and eradicate all other forms of life. This could be fatal in terms of the survival of species. All species, including man, are interdependent. Ecologists were teaching that one species cannot be destroyed without radically affecting others" (McCoy 1970: 154–155).

In the effort to preserve the wild, pristine parts of public lands in their original condition, conservationists often referred to the significance of such protected areas for wildlife. As Carl Buchheister, president of the Audubon Society, wrote in his organization's magazine in 1961, "In addition to the other powerful arguments for its enactment, the Wilderness Bill now before Congress is an important and necessary measure for the conservation of valuable wildlife" (Buchheister 1961: 153). Buchheister then enumerated the endangered species who "must have wilderness to survive." The bill did not pass that year but finally did become law in 1964, and a national Wilderness Preservation System for federal lands was established.

In terms of administrative policies, one of the more significant government reports to be published was produced by a committee chaired by A. Starker Leopold, who had followed in the footsteps of his famous father to become a nationally respected wildlife biologist. Leopold's committee, officially called the Advisory Board on Wildlife Management for the Secretary of the Interior, completed three reports, the first on big-game management in the national parks and the third on the national wildlife refuge system. But it was the second report, "Predator and Rodent Control in the United States" (Leopold et al. 1964), that drew the most attention, likely because of its controversial and emotional subject matter and also because Leopold and other prominent wildlife specialists made some strong statements in the report about both the PARC program and a societal change in attitude that demanded new practices from the existing corps of federal wildlife managers.

The Leopold committee opened the report with a historical recollection of a time when different values (or the absence of certain values) were characteristic of the American population: "In a frontier community, animal life is cheap and held in low esteem. Thus it was that a frontiersman would shoot a bison for its tongue or an eagle for amusement. In America we inherited a particularly prejudiced and unsympathetic view of animals that may at times be dangerous or troublesome" (Leopold et al. 1964: 28). The committee painted a picture of "conquest" and "incessant war" on large predators in particular, adding that "the maxim still persists that the only good varmint is a dead one." There was a need, apparently, to recognize our uncivilized roots and the prevalence of the negativistic value that had been formed in a time different from the present day. The committee continued: "But times and social values change. As our culture became more sophisticated and more urbanized, wild animals began to assume recreational significance at which the pioneer would have scoffed. Americans by the millions swarm out of the cities on vacation seeking a refreshing taste of the wilderness, of which animal life is the living manifestation" (Leopold et al. 1964: 28). Predators, the committee claimed, were an essential element of this wilderness, the wilderness that Americans wished to preserve. In addition, the primary targets of the PARC program—wolves, mountain lions, grizzlies, and even the reviled coyote—seemed to hold a special place in the cultural psyche of America. "The large carnivores in particular are objects of fascination to most Americans, and for every person whose sheep may be molested by a coyote there are perhaps a thousand others who would thrill to hear a coyote chorus in the night. Control programs generally fail to cope with this sliding scale of values" (Leopold et al. 1964: 28–29).

The committee did recognize that animal control was a complex problem, and that it was likely hard to avoid some degree of local population control of "troublesome" individuals that were causing damage or endangering human safety. But the central message was clear: "It is the unanimous opinion of this Board that control as actually practiced today is considerably in excess of the amount that can be justified in terms of total public interest" (Leopold et al. 1964: 29). In addition, in making their recommendations, the Advisory Board followed certain "tenets." The first was a statement quoted often by wildlife interests for years to come, thus constituting a significant declaration in the history of animal protection in this country: "All native animals are resources of inherent interest and value to the people of the United States. Basic government policy therefore should be one of husbandry of all forms of wildlife" (Leopold et al. 1964: 29). The Advisory Board, in its official capacity, had declared an end to the war on predators and more generally had finally dispensed with the idea that some animals were "good" and others "bad." And although it took several years before some of the more drastic recommendations would be implemented, overall the Leopold report had a major impact on the management of wildlife, particularly large predators. It was no longer just a few select species that the government had a duty to protect. Attitudes had evolved such that all native animals were considered to have some value. By highlighting this fact, the Leopold report served as the herald for a development that had been percolating through American society for many years previous to the report's publication.

International Interest in Species Protection

The belief that all native animals are significant was perhaps most strongly expressed in the rapidly expanding efforts to protect endangered species. This interest was not confined solely to the United States. Internationally in the 1950s and 1960s, a strong movement for the preservation of the world's fauna was developing. This movement was led by the International Union for the Conservation of Nature and Natural Resources (IUCN). The organization had been around in earlier incarnations since 1934 but became the IUCN in 1956 as its scope and influence began to widen (see chapter 5). The union's conservation research and assignments were delegated to six commissions, one of which was called the Survival Service Commission, whose primary responsibility was "to collect data on, and maintain lists of, all wild animals and plants that may be in danger of extinction, and to initiate action to prevent it" (Fisher, Simon, and Vincent 1969: 10). The lists evolved into the Red Data Books, officially published in 1966 in a large loose-leaf format meant to facilitate the insertion of updates as more information became available on the earth's troubled wildlife. The Red Data Books quickly came to be considered the authoritative source for all interested in the status of endangered species.

But the IUCN, whose main purpose was to serve as a consultant for United Nations agencies and governments, had neither the legal status nor the resources to raise the money necessary to fund many of the projects that they prescribed. It was apparent to some prominent conservationists that many species needed immediate help and that a quick influx of cash to the appropriate sources was the only way to initiate action. One such conservationist was British biologist Julian Huxley, who after returning in 1960 from a trip to East Africa wrote three dramatic articles for the London newspaper the *Observer* describing the disturbing decline of African wildlife and habitats. The articles generated a groundswell of concern in England. One businessman, Victor Stolan, saw an opportunity that he conveyed to Huxley in a letter. "If what is left in Africa and elsewhere is to be saved," Stolan wrote, "a blunt and ruthless demand must be made to those who, with their riches can build a shining monument in history" (quoted in Crowe 1970: 12). Under the guidance of Max Nicholson, director general of Britain's Nature Conservancy, and ornithologist Peter Scott, vice president of the IUCN, the World Wildlife Fund (WWF) came into being in 1961 as a tax-exempt charitable institution. WWF set up its headquarters in the same building in Morges, Switzerland, that housed the IUCN, and the two organizations worked closely together to advise and fund wildlife conservation projects around the world. As an international organization, WWF decided it could raise money more efficiently if it established "national appeals" in different countries. Britain (because of its connections with the founders of WWF) was targeted first in 1961, and the WWF branch in the United States was formed in 1962 (Crowe 1970: 13). In its first three years, WWF raised nearly two million U.S. dollars, much of it coming from individual donors. Not surprisingly, one of the first grant recipients was the IUCN. By 1968, 275 projects in 58 different countries had received support from WWF. The success of the organization both internationally and in the United States reflected the growing sympathy for endangered animals and the alarm over the degradation of the global environment.

The Endangered Species Acts of 1966, 1969, and 1973

The concern for endangered indigenous species in the United States had been an undercurrent for several decades, and a widening array of animals were drawing public attention. While international species such as the tiger, rhinoceros, and elephant enjoyed a celebrity status that was effectively employed by international conservation organizations like WWF and IUCN for raising money, American conservationists working to protect native species relied on appeals to preserve the cultural and ecological values associated with animals and plants. With more articles on "vanishing species" appearing in environmental magazines and other widely circulating periodicals, and with strong advocates in both scientific and political arenas, the movement to protect what was left of the native flora and fauna in the United States was steadily gaining momentum.

In 1964, the Department of the Interior established a Committee on Rare and Endangered Wildlife Species, consisting of nine biologists who immediately published their own "Redbook." It was the first federal list of endangered species in the United States, identifying sixty-three vertebrate animals that the committee members had determined were in danger of extinction (Yaffee 1982: 35). Later in the same year, Congress passed the Land and Water Conservation Fund Act. Although its general purpose was for "preserving, developing, and assuring accessibility to all citizens . . . of outdoor recreation resources," there was a section that specifically allowed funds derived from the act to be used to acquire "any national area which may be authorized for the preservation of species of fish and wildlife that are threatened with extinction" (Bean and Rowland 1997: 230). This provision granted the Department of the Interior "the authority to purchase habitat for endangered species preservation that for the first time was not on a species-by-species basis" (Yaffee 1982: 37). Up to this point, the government had responded ad hoc when a species was considered in danger of extinction, either through passing special legislation, as with the bald eagle, or in setting aside a refuge specifically for an endangered animal, as with the Aransas National Wildlife Refuge for the whooping crane. Congress now seemed willing to consider a more comprehensive approach to the problem of vanishing species.

As a result of internal pressure from the biologists in the Bureau of Sport Fisheries and Wildlife, external pressure from a public constituency increasingly concerned with protecting nongame animals, and recognition by congressional representatives of an issue that seemed to have political advantages in its growing appeal, Congress passed the Endangered Species Preservation Act of 1966 (Yaffee 1982: 39). The law had several notable features. First, its primary purpose was to establish a national system to coordinate the country's National Wildlife Refuges. While land had previously been set aside for wildlife in piecemeal fashion, the 1966 law provided not only organizational guidelines but also further authority to appropriate money for purchasing key habitat areas. The law also delineated the criteria for determining whether a species deserved federal protection. There was no mention of economic factors (i.e., the monetary value of a species); endangerment was to be a function of biological status, not the "worth" of the species to human beings.

But the 1966 act was also perceived as having significant weaknesses. Although conceived as "comprehensive," the law in fact only applied to "selected species of native fish and wildlife." It was evident from the Redbook list that only vertebrate animals would be eligible for protection. In addition, there were no measures to protect foreign endangered wildlife. Finally, there was a limited provision on the killing or harming of endangered species, which was prohibited only on federal lands. Otherwise such prohibition was left to the state governments. In general, while the 1966 act was an important first step, it came to be viewed as little more than a housekeeping bill for the Department of the Interior. It largely codified the powers the department had previously been employing and provided guidelines for standardizing the habitat acquisition program. But not long after the law was passed, those committed to stronger protection began to work on new legislation.

With the international dimensions of endangered wildlife fast gaining recognition in the United States through the work of international groups such as WWF and national groups such as the National Wildlife Federation, proponents for a more comprehensive law seized the opportunity to push for significant changes to the 1966 act. The result was the Endangered Species Conservation Act of 1969, the most notable characteristic of which was that it gave authority to the Department of the Interior to publish a list of wildlife "threatened with worldwide extinction" and to prohibit importation (Bean and Rowland 1997: 321). In addition, the act directed the United States to "seek the convening of an international ministerial meeting" at which delegates would negotiate "a binding international convention on the conservation of endangered species" (Bean and Rowland 1997: 324). As with the 1966 act, legislation to help struggling animal species was popular with both Congress and the public. Now the endangered species issue in the United States was officially extended to species living outside our borders, such as pandas, leopards, and cheetahs.

Another modification brought by the 1969 act was the expansion of the types of wildlife subject to protection. Although not often viewed as a significant legal change, the new definition is evidence of the incremental broadening of perspective that was underway. The 1969 act amended the 1966 act by redefining "fish and wildlife" as "any mammal, fish, wild bird, amphibian, reptile, mollusk, or crustacean" (Bean and Rowland 1997: 197). No longer was special consideration reserved for warm, furry, or feathered vertebrates or for sport fish. Largely, this amendment was included as a way to protect the American alligator from poachers who wanted the commercially valuable skins. Not only were the alligator and various nonmammal, nonvertebrate creatures protected under the new law, but the 1969 act also amended the Lacey Act to include these species in its prohibition on interstate and foreign commerce.

The expansion of the definition of wildlife to be protected was viewed as uncontroversial, and the bill sailed through the initial legislative stages, including the House vote, with ease. But before the bill reached the floor of the Senate, the fur industry learned of the expanded prohibitive features and tried to influence the decision-making process. The international dimension and subsequent effect on commerce would certainly have a marked impact on their business. While the law still passed eventually, it included several minor compromises, and although the fur industry was not an interest that engendered much sympathy, legislators and other supporters received their first taste of opposition on a subject that most had assumed was a "motherhood and apple pie" issue. The result was a law less prohibitive than originally conceived, although still significantly more potent than its 1966 predecessor (Yaffee 1982: 44–47).

Over the years leading to the final 1973 Endangered Species Act (ESA), the general wave of environmentalism sweeping the country gained momentum. The first Earth Day was celebrated in 1970. Congress passed numerous environmental laws, including the National Environmental Policy Act, the Clean Air Act, and the Federal Water Pollution Control Act. More types of people were becoming concerned about pollution and environmental quality. On the wildlife conservation front, the international convention called for in the 1969 act was finally held in March 1973. The Convention on International Trade in Endangered Species of Wild Fauna and Flora (CITES) added a new concept to the efforts to preserve rare species: the concept of differing degrees of endangerment. From now on, species "threatened" with extinction would also be deserving of protection—an idea adopted by those pushing for revisions to the 1969 act. Another important wildlife law, the Marine Mammal Protection Act (MMPA), passed in 1972, also contained the concept of "threatened" species, as well as the idea that subspecies and distinct populations of a species could be considered endangered.

Most notably, the promising comprehensiveness of the 1966 and 1969 acts was finally fulfilled in the 1973 act. Taking the lead of the MMPA, the 1973 act made it unlawful to "take" an endangered or threatened species anywhere in the United States, not just on federal land. Additionally, to "take" was defined very broadly as "to harass, harm, pursue, hunt, shoot, wound, kill, trap, capture, or collect." This was a bold move for the federal government, which previously had deferred to the states' traditional jurisdiction over resident wildlife. Although a small compromise was made allowing the states to adopt their own rules within federal parameters, the federal government was now undeniably making the rules.

But those federal "rules" also applied in-house: all federal agencies were required to make sure that their activities would in no way jeopardize the continued existence of endangered species. No longer, as with the earlier laws, could an agency use discretion in making trade-offs that would impact an endangered species. The directive applied to the species' habitats as well as to the organisms themselves. This element of the new law effectively made the entire federal government complicit in efforts to protect endangered species.

The most significant change, however, came once again in the definition of the wildlife that would come under protection of the law. While the 1969 act extended protection beyond mammals and birds to include reptiles, amphibians, mollusks, and crustaceans, the 1973 act completely threw out the categories and declared that all phyla of plants and animals were deserving of preservation. This meant that a rare butterfly species or an endangered flower enjoyed as much protection (theoretically) as a whooping crane or an Asiatic tiger. Not only had the federal government effectively co-opted the traditional right of the states to set their own wildlife management policies; it had done so in dramatic fashion by extending protection to those species that had previously been ignored in the laws of the land.

In the opening section of the 1973 ESA, the government enumerated the values that endangered species represented. To protect them was to recognize that such plants and wildlife were of "esthetic, ecological, educational, historical, recreational, and scientific value to the Nation and its people" (U.S. Congress 1973: sec. 2 [a] [3]). This official statement was powerful not only for its inclusiveness, but also for the notable absence of any reference to economic or utilitarian value. The intangible values that had been expounded in articles and books and implied in previous laws were finally clearly spelled out and recognized as worthy.

Remarkably, the passage of the law was uncontroversial. One might have thought that the potential restrictions and the elusive quality of the values stated in support of the act would have caused an outcry of opposition. But beyond some minor bickering over the federal juris-

diction issue and the inclusion of plants as protected species, the 1973
ESA easily passed through both early and later stages of the legislative
process. Indeed, the final Senate vote (92–0) and House vote (390–12)
suggest that it was precisely the intangible and symbolic nature of the
values at stake that had convinced lawmakers and potentially affected
interests that this law could do little or no harm. The fur industry was
the object of scathing attacks from animal defenders and was suffering
badly in the arena of public opinion; its representatives already had
their say in the deliberations for the 1969 act. It was unclear who else
might be adversely impacted by this law, except poachers, miscreants,
and government agencies (the last of which were directed simply to
monitor their own activities). As the values listed in the ESA suggested,
this law—even in 1973—was perceived as a recognition of a change in
the American attitude toward its wildlife. More inclusive in terms of
what organisms were considered valuable and why, the 1973 ESA was
viewed by legislators as a prohibitive act that would serve simply to
redirect emphasis of development and management programs. As
such, they believed, it would cost little, either monetarily or politically
(Yaffee 1982). No one foresaw the controversies and numerous cases
that would ensue.

One of the primary effects of the court cases besides clarifying the
broad language of the law was to interject economic value back into
the mix of considerations when addressing impacts on endangered
species. The most famous case associated with the ESA was *Tennessee
Valley Authority [TVA] v. Hill.* The TVA in 1976 was preparing to
close the gates on the Tellico Dam, a project that had begun before passage of the 1973 act. The action, according to the plaintiffs in the case,
would destroy the habitat and thus the only known population of the
recently discovered, endangered snail darter, a two-inch-long fish that
to the untrained eye resembled a minnow. At the time the snail darter
was discovered, the TVA had already constructed major portions of
the dam. As such, there was some question as to whether the ESA applied retroactively. Millions of dollars of expended funds and additional losses were at stake. The language of the law seemed clear: economic value was trumped by all of the other values listed in the ESA.
But this was a conflict of proportions—both in big money and small
fish—that was unforeseen by legislators in 1973.

In 1978, the case was heard by the Supreme Court, and in a 6–3 decision the Court ruled that the ESA clearly prohibited the TVA from completing the dam and ostensibly ending the snail darter's existence. But in the same year as the TVA decision, Congress amended the law to include a formal process whereby a select committee would consider exemptions. Dubbed the "God squad" by disgruntled environmental groups, the committee essentially had the power to decide whether or not the benefits of a project lacking alternatives outweighed the values associated with the existence of a species. In the end, although the TVA case had inspired the amendment, the Tellico Dam was denied exemption to the ESA by a unanimous decision of the committee.

But the potential for conflicting values had been starkly revealed in *TVA v. Hill*. It was apparent that protecting endangered species, especially when considering all phyla of plants and animals, was far more complicated than the 1973 Congress had assumed. Preserving familiar, charismatic, relatively visible species like the whooping crane, the timber wolf, and the white rhino was one thing. But giving protection to a tiny fish whose value was unclear to many American citizens was quite another. Did the snail darter really hold any aesthetic, ecological, historical, recreational, or scientific value? What harm would it have done to extirpate the species? Would it have been missed? In this way, the snail darter case tested the validity of the ESA and the resolve of its proponents. But most significant, the decision to protect the fish was an affirmation of the growing principle that the entire diversity of species on earth was important to maintain. Even when the value of a species was unclear, the simple fact that it was a discernible unique unit in the hierarchy of the living world gave it a right to existence that was supported by the multiple values now acknowledged. The snail darter episode brought into focus more than ever this attitudinal change implicit in the 1973 ESA.

By the 1970s, numerous environmental groups had come into existence, many of which were dedicated to preserving the endangered animal and plant species that could be protected under the Endangered Species Act. The most popular of these groups were often wildlife organizations that had been around for many years and had developed more inclusive perspectives in parallel with the societal changes in environmental values. Groups like the National Wildlife Federation, National Audubon Society, Defenders of Wildlife, and the World Wildlife Fund enjoyed surges in membership and an influx of resources that in earlier years would have been unimaginable. This expansion in support can be attributed largely to the public's general interest in environmental issues but also more specifically to the high-profile cases of endangered species used by the organizations in soliciting members. In short, endangered species served as the symbols of a fragile, broken natural world that needed immediate protection from humans. The values listed in the 1973 Endangered Species Act, plus the ethical ramifications of extinguishing unique forms of life, indicated that all living organisms needed to be preserved. In this context the idea of maintaining diversity in the natural world began to take hold.

Conclusion

The concern for endangered species is not exactly the same as a concern for maintaining a diversity of species. There was a dramatic element of crisis that surrounded species in danger of extinction, while the value of diversity lay in benefits accrued from a healthy-functioning, complete biota. But protecting endangered species was obviously an essential part of preserving diversity; the movement toward seeking protection for all species, whether endangered or not, was the obvious next step in wildlife conservation. The desire to maintain diversity reflected all of the values identified with protecting endangered species. Diversity in an ecosystem provided ecological value theoretically by maintaining a more stable foundation for natural processes and functions. A full complement of flora and fauna was more aesthetically pleasing to humans and provided more recreational opportunities. In addition, protecting diversity was a way of protecting opportunities for scientific investigation, safeguarding potential utilitarian benefits that could translate into vast commercial value, satisfying the humane impulse of those who sympathized with the nonhuman world, and recognizing and maintaining the ethical obligations of humans to other organisms.

But while the concern over extinction has become the front line of the war for maintaining species diversity, it is important to remember the values that built the conservation traditions of our society and to discern what these values imply. The early emphasis on utilitarian values—commercial and economic—illustrates the human dependence on living resources for life-sustaining goods. The rise of sport and recreational value tells of our need to engage the natural world and its physicality. The scientific value of protecting various species reflects our need for knowledge gained from studies of animals. Humanitarian values remind us not to discount the motivating force of human sympathy in conservation. Conversely, the negative value historically assigned to predators illustrates the power of fear and hatred in our relationships with certain organisms. The investment of cultural value within the natural world shows how we use animals to represent the national and societal values that we hold dear. The poetic expression of the aesthetic value of species helps to broaden our notions of which animals and plants are worth protecting. The recognition of ecological value and the interconnectedness of the natural world solidifies the view that all species have a specific role to play in the workings of nature. Finally, the effort to protect endangered species highlights the ethical imperative that we try not to let any animal or plant become extinct.

These trends in the history of concern for species provided the medium that allowed the concept of biological diversity to flourish. The idea of using *diversity* as the umbrella term for all species and the values that they provide became popular in the literature of the 1970s. But species conservation is only one part of the multilayered framework outlined in biological diversity. To complete the picture, we must also examine the history of the concern for genes and ecosystems.

4 The Concern for Genetic Diversity
Science, Agriculture, and Conservation Biology

The concern for genetic diversity, like the concern for species diversity, has multiple sources in American society. Just as the conservation of the variety of species was important to a number of constituent groups in the twentieth century, the conservation of certain "gene pools" or sources of "germplasm" was vital to diverse interests. This chapter examines three broadly defined groups whose interest in the genetic variety in the natural world fueled conservation efforts. First, the scientific community, in its efforts to understand the mechanisms of inheritance (and thus the engines of evolution), was particularly focused on the gene and its variation after the rediscovery in 1900 of Gregor Mendel's work. Second, agricultural interests conducted their own scientific research in efforts to improve crops by employing the genetic resources both of old cultivated plants and of wild relatives of our present-day crops. Third, scientists concerned with saving species from extinction—those who would come to be called "conservation biologists"—used their knowledge of genetics to evaluate and manage populations and species that were in danger of extinction.

The range of values placed on maintaining a variety of genes evidently was not as wide as that placed on protecting species, but it is still apparent that a number of different values came into play in the efforts to protect genetic diversity. For example, the scientists working on solving the mysteries of the gene, while not an overt force in pushing for conservation in the natural world, still expressed the scientific value of the natural variation that occurred in genes. In addition, the milieu of intellectual excitement over discoveries in genetics certainly played a role in adding to the importance of maintaining genetic diversity. The agricultural interests, which benefited greatly from the scientific advances, were the first to call attention to the dangers of losing the genetic diversity in nature. Agriculture ostensibly valued genetic variety above all for utilitarian purposes. However, most of the early calls of concern from this interest also mentioned the importance of learning about the evolution of certain crops and domesticated animals, and it is apparent that those concerned with losing genetic variability were also worried about losing material of both scientific and cultural value. Finally, with the rise of conservation biology in the late 1970s, a community of conservation-minded scientists had decided to employ their knowledge of genetics as one of many tools in their efforts to protect a much wider range of values. Here the concern for genetic diversity blends with the concern for species diversity, and all the values associated with conserving species—including utilitarian, aesthetic, ecological, and ethical values—were apparent in the efforts of conservation biologists to protect genetic variation.

It is sometimes difficult to separate out the concern for genetic diversity from the concern for species diversity. The most obvious way to protect genes is to protect the species that carry them. But it is evident in the language and intention of those concerned with maintaining genetic diversity that the molecular level of the gene is the foundation of their perception of conservation. The most utilitarian perspectives view a species simply as a vessel for the real locus of value, genetic material. But even some concerned with protecting species choose to focus on the genetic level because they believe such a perspective reveals the fitness of a population for survival. Within the paradigm of interdependency in nature, it is apparent that we cannot effectively protect species without maintaining genetic diversity, and likewise we cannot protect the global gene pool without preserving species diversity. Equivalent statements can be made regarding ecosystem diversity. But genetic diversity represents a special challenge: as Margery Oldfield points out, "Because we cannot immediately see and touch genetic materials and because their biological sources and economic uses are often obscure to us, it is difficult to discern the essential role they play in sustaining our lives and societies" (Oldfield 1984: v). Many individuals and organizations have worked tirelessly to communicate the importance of protecting genetic diversity. The place of genetic diversity within the framework provided by the concept of biological diversity is no accident—it was brought about by various sources in the twentieth century that expressed their concern.

The Scientific Interest
Discovering Mendel

The study of genetics and the molecular mechanisms of diversity has dominated the science of biology for the past century. When the scientific community "rediscovered" Mendel's work in 1900, the flurry of experimental activity and postulating of theories was without doubt one of the most exciting times in the history of science. Many accounts of the advances in genetics have been written. Most have used superlatives in trying to describe the significance of the breakthroughs in genetics for understanding life on earth. As John Moore wrote, "The science of biology has few stories more interesting than the history of man's attempts to explain heredity and development." He asserted that the discovery in 1944 of the chemical substance of which genes are composed is matched "in importance only by the theory of evolution" (J. Moore 1963: ii). L. C. Dunn similarly observed that "the biological science which has undergone the most rapid development in the first half of the twentieth century and which has most profoundly affected the development of biology as a whole is undoubtedly genetics" (Dunn 1965: ix). Ernst Mayr declared that an understanding of genetics is the core of all modern biological thought: "A comprehension of the fundamental principles of inheritance is a prerequisite for a full understanding of virtually all phenomena in all other branches of biology, whether physiological, developmental, or evolutionary biology" (Mayr 1982: 629–630). Indeed, it seems that only the splitting of the atom rivaled the advances in genetics for the most important scientific discovery in the twentieth century.

The first decade of the century was dedicated to exploring, testing, and refining the observations that Mendel had made about thirty-five years earlier in his carefully recorded experiments with pea plants. The rediscovery of Mendel's work is most often credited to three scientists working independently—Hugo de Vries, Erich von Tschermak, and Carl Correns—who apparently came upon Mendel's original 1866 paper while searching for precedents that might help them in their own studies (Witt 1985: 121). All three were botanists. Both de Vries and Correns had already published results of systematic crossing experiments that they had begun in 1892, while von Tschermak apparently played a less active role in the initial discovery but was instrumental in "directing the attention of the plant breeders to the importance of Mendelian genetics" (Mayr 1982: 730). The major scientific contribution of Mendel's work, verified by those who followed, was the first rule of genetics: each heritable trait (like the pigment color of the pod on the pea plants that Mendel studied) is represented in the fertilized egg by two (and only two) factors—one coming from the father and the other from the mother (Mayr 1982: 721). Mendel's focus on successive hybrid generations originating from plants that exhibited two different characters of a single trait also led him to theorize about the dominant and recessive nature of the appearance of certain characteristics (Dunn 1965: 7). The existence of these "factors," as Mendel called them, the different expressions of the same characteristic, one paternal and one maternal, was the foundational idea upon which the science of genetics was built.

The term *genetics* did not exist until William Bateson suggested it in a 1907 article as a title for the rapidly evolving science of inheritance (Whitehouse 1973: 449). Interestingly, the term *gene,* proposed in 1909 by Danish scientist W. L. Johannsen, was not named after Bateson's new moniker, but instead was derived from de Vries's *pangens,* his term for Mendel's "factors" (Mayr 1982: 736).

Focusing on Diversity

It was diversity—the inherent variation of the genetic constitution of species—that first triggered the new ideas about the mechanics of inheritance. For Mendel, it was the simple variation in pea plants—their color, height, and seed texture—that stimulated his thoughts on how physical traits were passed from generation to generation. Mendel's focused experiments, however, never went beyond the assumption that only two different expressions of a particular trait were available in the gene pool of a species. But soon after Mendel's work had gained the attention of the biological sciences, a French biologist named Lucien Cuénot began experiments that would confirm that "the number of alternatives for a particular character which showed Mendelian inheritance was not limited to two" (Whitehouse 1973: 137). Working with mice, Cuénot found that crossing individuals of three different coat colors revealed that the three characters (yellow, gray, or black fur) behaved in a Mendelian fashion; there were simply more possibilities of expression, more ingredients to impact the new generations. This initial discovery of multiple "alleles" (originally named "allelomorphs" by Bateson in 1902 and later changed to alleles) led to other research that would further elucidate the behavior of crossed expressions of a certain gene. This research included experiments with wheat and oats by Nilsson-Ehle (1908, 1909), corn by Emerson (1911), and rats by Castle (1914) (Sturtevant 1965: 60). One of the most well-known examples of research on multiple alleles was Sturtevant's work on human blood types (1913). Based on his results, Sturtevant theorized that different alleles were "alternate states of the same gene (locus)" (Mayr 1982: 754). But Mendel's basic law still held firm in these experiments. Only one allele was represented in a given reproductive cell of an individual, but it was able to combine with a different allele in the reproductive cell of another individual. The early experiments on multiple alleles confirmed that the opportunities for different combinations were numerous.

As the theory of the gene was elaborated, the questions about its internal workings seemed innumerable. In particular, what was the source of multiple alleles? How did different individuals acquire different expressions of the same gene? The existence of this variation was an important component of Darwin's theory of evolution and natural selection, but Darwin himself had no idea how the diversity in nature had come about. After a decade of picking apart Mendel's work and the experiments and theories it inspired, the scientific community was ready to move to a new level of understanding. One scientist, T. H. Morgan, decided to study genes that had "changed" in unexpected ways when being passed from one generation to the next. These "mutations" became the centerpiece for the next major advance in genetics (Mayr 1982: 738).

As Mayr points out, "The year 1910 is almost as famous in the history of genetics as the year 1900; it was the year of Morgan's first *Drosophila* publication" (Mayr 1982: 744). In 1909, Morgan had switched from experiments with mice and rats to experiments with *Drosophila* fruit flies, subjects that required less care and had a much quicker turnover rate for successive generations. When a single white-eyed male fly appeared in what should have been a generation of red-eyed flies, Morgan knew he had stumbled on a remarkable opportunity. The resulting experiments confirmed the existence of linked genetic traits and the occurrence of characteristics that cross over, and they contributed significantly to clarifying the mechanisms of variation and heredity (Mayr 1982: 752–753).

It is not an overstatement, therefore, to claim that it was the diversity within species—the natural variation in the expression of genetic traits—that stimulated the great discoveries of biologists studying genetics. Perhaps this seems like an obvious point, but it is important for assessing the significance that genetic diversity is perceived as possessing. The scientists working on the problems of inheritance greatly valued the existence of the internal diversity of species. At the 1932 Genetics Congress held at Cornell University, there was "the realization that without coordinated maintenance, genes (mutants) could be lost through neglect," and in order to share information about efforts to protect against such losses, the scientists performing experiments with corn started the Maize Genetics Cooperative Newsletter (Wilkes 1983: 146). In the late 1930s, those working with *Drosophila* and maize came together to establish their own "stock centers," such that scientists had access to any particular "mutation" that they might need for their experiments (Landauer 1945: 489). Although one might argue that this expression of scientific value had little to do with the concern for natural genetic diversity, it seems apparent that the overall atmosphere of discovery surrounding the science of genetics conveyed a sense of importance that contributed to the desire to protect genetic diversity. The study of genetics has given us insights into the existence and evolution of life like no other branch of science before it, and the genetic diversity that inspired this research illustrates the great scientific value inherent in the components of the natural world.

The Agricultural Interest in Genetic Diversity
The Nineteenth Century: Plant Importation and Introduction in the United States

The applications of these seminal scientific discoveries are more tangible manifestations of the value of genetic diversity. Previous to the advances in the study of genetics, there was great interest, especially in the United States, in the diversity of "useful" plants that had already been manipulated and domesticated through centuries of human intervention. Because the European settlers had found the native agricultural resources in the New World to be lacking, there was a constant effort to obtain seeds of new crop varieties from faraway lands. The importation and introduction of bountiful exotics was a main priority. The literature on American agriculture often quotes the following words of Thomas Jefferson: "The greatest service which can be rendered to any country is to add a useful plant to its culture" (in Hyland 1984: 6).

Jefferson's high opinion of plant introduction apparently was shared by many. Officials from the United States who traveled abroad often sent or brought back seeds of interesting crop species. In 1770, Benjamin Franklin sent the first soybeans back to Pennsylvania from Europe (Witt 1985: 120). Jefferson himself, when visiting Italy, secretly brought home upland rice with the intention of introducing it in South Carolina (Fowler and Mooney 1990: 40). The interest of the American agricultural community in obtaining new varieties was so pervasive that numerous agriculture "societies" were established in the early nineteenth century largely for purposes of sharing seeds and information. Some of these societies even petitioned those who traveled abroad for specific plants and crop materials. Elkanah Watson, founder of the Berkshire Agricultural Society, used his prominent status as a merchant and banker to "systematically request seeds from American consuls all over the world" (Kloppenburg 1988: 52–53).

In 1819, the practice of sending home seeds started to become institutionalized when Secretary of the Treasury William Crawford sent a letter to all American diplomats and naval personnel "directing them to send useful seeds to Washington for distribution to farmers and anyone else who wanted them" (Witt 1985: 29). At first it was an informal program, and there was no established infrastructure for receiving and assessing the new seeds. Eventually, the task of cataloguing and distributing the seeds fell upon Henry Ellsworth, the commissioner of patents. Because of Ellsworth's position, this move has some significance for the history of plant breeders' rights and the future patent protection of certain crop varieties. But more important, it was Ellsworth's strong support of the seed distribution program that allowed it to flourish in the 1820s and 1830s. Ellsworth owned large tracts of land in the Midwest, and it seems likely that he had some personal interest in the discovery and importation of economically useful plant species. Whatever his motivation, his efforts led in 1839 to the congressional appropriation of funds to pay for an official program in the collection and distribution of foreign seeds and plants (Kloppenburg 1988: 55).

By the time President Lincoln established the U.S. Department of Agriculture (USDA) in 1862, the Patents Office had overseen the distribution of more than one million seed packages. The new agriculture department took over the responsibility of doling out the free seeds, and the distribution activities continued until 1925 (Witt 1985: 30). But it was not until 1898 that the USDA created a special division, the Office of Seed and Plant Introduction, responsible for finding and collecting useful plants (Hyland 1984: 6–7). It seems that the new office was established in response to complaints from the newly emerging American seed industry. Fifteen years before, the American Seed Trade Association (ASTA) had come into being and had immediately directed its energies toward curtailing the distribution of free seed and passing legislation that would encourage private investment in crop development (Pistorius and van Wijk 1999: 32). Their charges against the federal government centered on the lack of "quality control" in the USDA's distribution program. The ASTA claimed that many of the free seeds were inferior varieties that should not displace crops already established in the United States (Witt 1985: 31). With these criticisms in mind, the USDA set out to explore and collect plants from around the world more systematically. The early explorers hired by the USDA contributed a great deal of knowledge about different species and varieties of plants and collected vast amounts of plant specimens in the first decades of the twentieth century.

The Advent of Genetics and New Ideas about Breeding

It was, of course, around this time that the rediscovery of Mendel's work had set the scientific community on fire. As important as this was to the science of biology, the implications for professional breeders were just as significant. One of the first to apply the principles of Mendelism to crop research was the British scientist R. H. Biffen. In his 1906 paper "Mendel's Laws of Inheritance in Wheat Breeding," Biffen showed how resistance to a specific disease was passed on according to Mendel's rules and controlled by individual "factors of inheritance," physical components that would later be called "genes" (Pistorius and van Wijk 1999: 36). With this new theory of inheritance now operational in an agricultural example, plant breeders had been given a new vision for selecting varieties to improve crops. Instead of looking at the entire plant, breeders now needed only to focus on individual genetic traits. Freed from the labors of selective breeding, the new agronomists worked with hybridizing techniques such as "backcrossing" to produce varieties with predictable characteristics. This change in approach, writes one historian, was "a watershed in the development of plant breeding in the United States. No longer was the breeder's task to adapt elite germplasm from other countries to American conditions, it was now to improve established varieties by incorporating particular exotic characters" (Kloppenburg 1988: 80).

While the Europeans, particularly the British, first took the lead in scientific research, it was the agronomists from the United States who were more successful in applying genetics to crop development. At the turn of the twentieth century, Europe still maintained colonial control over many tropical countries and had unhindered access to the agricultural resources of these countries. For this reason, the European scientists tended to view Mendel's laws as a brilliant scientific discovery but their applications as unimportant. The United States, in contrast, was searching for ways to improve and strengthen its agricultural security, and therefore the American scientists paid more attention to the practical problems presented by crop development. It would take decades before the new knowledge produced any significant advances. But in the ensuing years, the USDA would build four federal Plant Introduction Stations designed to store and evaluate hundreds of thousands of plants and seeds to be collected. Such collections formed the foundation for any future agricultural research that would be undertaken in the United States (Pistorius and van Wijk 1999: 37–38).

The newfound knowledge of genetics, combined with the enthusiasm for plant exploration, produced not only numerous discoveries of new plant varieties but also important theories about the evolution of cultivated plants. For the first time, explorers and breeders began to see ancient landraces (old cultivars) and wild relatives of modern crops as potentially valuable specimens. Once thought to be useless, these plants were now examined by breeders for valuable traits that might be isolated using crossbreeding methods. The new interest in plants once ignored broadened the possibilities of what would be considered a valuable plant. For this reason, explorers were eager to learn which areas of the globe had particularly large numbers of species whose genetic materials were useful for breeding with modern crops. In 1882, Swiss botanist Alphonse de Candolle had published foundational studies on the origins and histories of cultivated plants (Witt 1985: 121). But it was a Russian scientist—Nikolai Vavilov—who would come up with the most influential theories of the world's "centers of diversity."

In the 1920s and 1930s, Vavilov, a geneticist by training, organized and led expeditions in over fifty countries on all of the continents (excluding Antarctica) and collected more than fifty thousand seed samples of crop plants and their relatives (Plucknett et al. 1987: 62). In his extensive journeys, he observed that some regions contained a great deal more diversity in crop species and their wild relatives than others. Vavilov theorized that these diversity-rich centers were likely where the ancient ancestors of modern crops first flourished and were coaxed by early humans into their present-day form. His reasoning was simple, as one botanist summarizes: "The place of origin of a species of cultivated plant is to be found in the area which contains the largest number of genetic varieties of this plant" (Zohary 1970: 35). Vavilov believed that high levels of variation meant that a plant had enjoyed a long evolutionary history, indicating that it had been present in the diverse regions longer than other regions. Hence, these "centers of diversity" were synonymous for Vavilov with "centers of origin." In his first assessment in 1926 he proposed six geographic centers for cultivated plants, and by 1935 he had revised his global map to show ten geographic centers (Zohary 1970: 35).

As later scientists modified the boundaries of his regions, Vavilov's theory of the centers of diversity as the original birthplace of certain plants came under some criticism. The role of topography (more mountainous regions provided more opportunities for variation) and the absence of wild relatives in some of his proposed centers of origin made it clear that high diversity did not necessarily mean that a plant had been evolving in a particular region for the longest time. However, Vavilov's maps formed the basis of all modern work on crop species evolution and represented international interest in the genetic diversity and evolution of cultivated plants.

Vavilov inspired a whole generation of explorers in their efforts to discover new sources of germplasm. It is important to note that these explorers were not just treasure hunters looking for new species. Focused on the gathering of raw material, the new strategy of collection, as illustrated in Vavilov's efforts, was to acquire as much diversity as possible. These men were trained botanists who were able to distinguish between closely related varieties of the same species. One objective was to collect those plants that had been overlooked. As Knowles Ryerson, a plant explorer for the USDA, wrote in 1933: "Species and varieties which in themselves have little or no intrinsic value become of first importance if they possess certain desirable characters which may be translated through breeding" (quoted in Kloppenburg 1988: 80). The new collection strategy was evident in the numbers of samples being catalogued in U.S. agricultural research laboratories. While the number of USDA expeditions remained the same from 1925 to 1930 as in previous years, the number of accessions per five-year period nearly doubled. "Plant breeders looked not so much for new introductions as for breeding material, not so much for a superior variety that might be adapted to American conditions but for a plant with perhaps only one superior characteristic. Hence, they collected a much broader range of germplasm" (Kloppenburg 1988: 80). While not all explorers were as fervent as Vavilov, who wished to "cover the globe" and collect the full range of variation present in the relatives of modern crops, the basic philosophy was the same. In short, the genetic materials of once "useless" living organisms were now considered valuable resources. Most significantly, breeders were no longer in search of just *organisms,* but were focused on *genes,* and above all on the *diversity* of genes that occurred in the natural world.

The Problem of Genetic Erosion

With this new interest in the variety of crop species' relatives came the recognition of a paradox facing the breeders and collectors. As new varieties were developed and sent out to farmers around the world, the fields where old cultivars and wild relatives had grown for many years might likely be plowed under and replanted with the new, improved crops. In the USDA *Yearbook of Agriculture, 1936,* H. V. Harlan and M. L. Martini first warned the agricultural world of this threat to germplasm resources. In their often-cited article on barley breeding, the authors poetically summarized the potential danger: "In the hinterlands of Asia there were probably barley fields when man was young. The progenies of these fields with all their surviving variations constitute the world's priceless reservoir of germplasm. It has waited through long centuries. Unfortunately, from the breeder's standpoint, it is now being imperiled. When new barleys replace those grown by farmers of Ethiopia or Tibet, the world will have lost something irreplaceable" (Harlan and Martini 1936: 303, quoted in Pistorius and van Wijk 1999: 65).

Here was perhaps the first statement of a dilemma that has disturbed breeders up to the present day. Garrison Wilkes (1983) describes the problem as "genetic erosion," when an area of diverse germplasm resources is replaced with a genetically "superior" monoculture. It seemed almost as if explorers collecting genetic variants were racing against a clock which they had set against themselves. The agricultural improvements they had instigated were becoming responsible for potential losses of diversity. The answer to the problem in the 1930s and 1940s was to collect and store as many samples as possible before it was too late. As Harlan and Martini wrote in the same article cited above: "The plant breeder has every reason to feel gratified and undoubtedly the time is not too far distant when the entire acreage will be planted to pure-line varieties. There is, however, one rather disconcerting problem raised by the plant breeder's success. In a way, we lose whatever we gain. . . . The breeder is helpless without living material of diverse character" (Harlan and Martini 1936: 315–316, quoted in Pistorius and van Wijk 1999: 65).

One solution, likely considered unacceptable to most breeders and agronomists, was noted by famed geographer Carl Sauer in 1941.

When the U.S. government and certain philanthropic organizations entertained plans to help Mexico increase its wheat and maize production, Sauer commented: "A good aggressive bunch of American agronomists and plant breeders could ruin the native resources for good and all by pushing their American stocks. And Mexican agriculture cannot be pointed toward standardization on a few commercial types without upsetting native culture and economy hopelessly. The example of Iowa is about the most dangerous of all for Mexico. Unless the Americans understand that, they'd better keep out of this country entirely" (in Kloppenburg 1988: 162). By highlighting the scientific, cultural, and potential utilitarian value of seemingly "inferior" species, Sauer, Harlan, and Martini hoped to convince the agricultural world that there were important benefits in protecting diversity that must be balanced with the value of increasing yields. But such warnings largely went unheeded. Most in the agriculture business felt that the benefits of replanting old fields with genetically superior crops outweighed the potential loss of less productive relatives.

In addition, the effort to collect a representative cross-section of the earth's diversity was underway, and some chose to bet on the idea that germplasm resources could be protected in the world's gene banks. But with this heavy emphasis on collection and storage from 1900 to 1930, the facilities available to carry out the mission were reaching capacity (Pistorius and van Wijk 1999: 65). By the time the economic downturn of the 1930s gripped American farmers, many of the USDA programs, including plant exploration and introduction, were severely curtailed (Hyland 1984: 10). Some in Congress even attacked the scientific breakthroughs in crop "efficiency" as contributing to hard times. In the early 1930s, economic problems of crop surplus production were blamed on the increase in yields that agricultural research had provided. The result was that Congress had an excuse to cut back heavily on funds for any further expansion of storage and research facilities (Kloppenburg 1988: 85).

But the antiscience sentiment did not last long. Secretary of Agriculture Henry A. Wallace (founder of the Pioneer Hi-Bred Company and later Roosevelt's vice president) was an indefatigable champion for scientific research and development. Through his own lobbying efforts and by publishing articles with titles like "Give Research a Chance" (1934), Wallace was able to convince Congress to pass the Bankhead-

Jones Act of 1935. Not only did this bill add significant funds to the USDA's research program, but it also called for the construction of nine regional centers and an expansion of the extension service (Kloppenburg 1988: 86). The law was the boost that agriculture in the United States needed to lift itself out of hard times. Although the regional centers would not get built for over ten years, the new funds gave scientists the opportunity to begin to evaluate the huge collections obtained in the previous decades and sort through the possible crop improvements that could be made.

The mid-1930s marked the time when the promise that Mendelian genetics had provided agriculture was finally to be realized. Breeding by selection was now almost completely supplanted with breeding by hybridization. Corn provided the most dramatic example. In 1933, less than 1 percent of agricultural land in the "corn belt" was planted with hybrid varieties; within ten years 78 percent of the same land was growing hybrid corn (Witt 1985: 121). Other crops, such as cotton, wheat, and soybeans began a dramatic increase in yields that would continue for many decades (Kloppenburg 1988: 89). The power of the genetic techniques that agronomists now wielded was undeniable. As G. Hambridge and E. Bressman wrote in their conclusion to the *Yearbook of Agriculture, 1936:* "Now the breeder tends rather to formulate an ideal in his mind and actually create something that meets it as nearly as possible by combining the genes from two or more organisms. . . . In this connection, he has a new confidence . . . he has a vision of creating organisms different from any now in existence, and perhaps with some remarkably valuable characters" (quoted in Kloppenburg 1988: 88).

National and International Interest in Germplasm
in the Postwar Years

Even with the promise of new methods of plant breeding, the interest in conserving germplasm in the wild had not gained much momentum. Protection still focused on either domestic stocks or on those varieties and organisms useful in empirical scientific research. As noted earlier, the geneticists working with *Drosophila* and maize had organized to establish a "pure culture depository of mutant stock" in the late 1930s. In 1940, the National Research Council's (NRC) Division of Biology and Agriculture held a conference dedicated to the maintenance of pure genetic strains (Wilkes 1983: 146). After a predictable decline in interest during the war years, Professor Walter Landauer, chairman of the NRC's Committee on the Maintenance of Pure Genetic Strains, published an article in *Science* entitled "Shall We Lose or Keep Our Plant and Animal Stocks?" urging both scientific and agricultural interests to take renewed steps to protect "the raw materials—the mutations— as they are provided by nature" (Landauer 1945: 497). It was apparent, however, that the author was concerned mostly with germplasm diversity already collected and preserved in storage labs. The characteristic of the concern around the 1930s and 1940s was technical—proponents of genetic diversity conservation worried about the care of stocks in storage and the government funding that would support the proper maintenance into the future.

On the domestic front, the USDA was reorganized by the 1946 Research and Marketing Act. Given renewed direction "to stimulate research and encourage the discovery, introduction, and breeding of new and useful agricultural crops, plants, and animals," the USDA set out to expand their capacity as they had been directed to do in the Bankhead-Jones Act of 1935 (Wilkes 1983: 147). Four regional Plant Introduction Stations were established over the next eight years, bringing the total number to eight (Hyland 1984: 11). These stations were used largely for evaluation, and in 1956 Congress finally set aside money for the National Seed Storage Laboratory (NSSL) to be built in Fort Collins, Colorado (Brown 1984: 32). Although funds were limited for many years after the NSSL was built, the combination of the expanded regional labs and the national storage center was to be the core of what the federal government would call its National Plant Germplasm System.

But at the next NRC meeting on plant and animal germplasm, a new international interest was introduced to the American agricultural community. The 1946 conference was attended not only by state and federal officials but also by representatives from a recently established United Nations agency, the Food and Agriculture Organization (FAO). The FAO, with its constitution ratified in 1945 by the requisite number of countries and its approved mandate "to achieve freedom from want for all peoples in all lands," was poised to become a significant international force in agriculture and the concern for genetic diversity as it related to crop species (FAO 1985: 10).

As the United States was taking basic steps for establishing a federally supported program, the international players began to act. In 1947, an FAO subcommittee recommended that the new UN agency use its resources to distribute information about genetic research and facilitate the "free exchange" of germplasm between countries wishing to strengthen their agricultural sectors (Witt 1985: 22). Aid right after the war focused for the most part on helping countries regain their production capacities, which had been disrupted and dislocated (FAO 1985: 17). By 1953, this goal had been achieved; the problem that arose in the 1950s was surplus. But in 1957, the FAO was able to focus some of its efforts once again on the issue of germplasm resources, and the organization published the first issue of the *Plant Introduction Newsletter.* The main purpose of this periodical was to inform readers where sources of a particular germplasm were located so that they might obtain them. In the publication of the sixth issue in 1959, the FAO compiled a world list of germplasm banks and their custodians (Wilkes 1983: 157). The germplasm distribution efforts were capped off in 1961 with the FAO-coordinated World Seed Year, which promoted the global use of "first-class seed of superior crop and tree varieties." Ostensibly, this campaign was designed to help developing countries replace their traditional seed stocks with modern, improved varieties (Pistorius and van Wijk 1999: 92).

At the same time that the FAO was expanding its reach into the plant germplasm world, another international force was also gaining strength and exerting influence. In 1943 (in apparent conflict with Carl Sauer's earlier warning), the Rockefeller Foundation and the Mexican government established the Mexican Agricultural Program in an effort to increase Mexico's wheat and corn yields and thus make the country less dependent on foreign imports. Fourteen years later Mexico was able to meet its own demands for the first time (Witt 1985: 122). Based on this success, in 1960 the Rockefeller Foundation next teamed with the Ford Foundation and the Philippine government to fund the first international agricultural research center—the International Rice Research Institute (IRRI). By 1964, a second international research center was built in Mexico, the International Center for the Improvement of Maize and Wheat (CIMMYT). In this same year, the FAO established its first Crop Research and Introduction Centre in Izmir, Turkey. These two international structures—one based on the global philosophy of the United Nations and the other driven by U.S.-based foundations with political (as well as humanitarian) interests—were competing to help the underdeveloped countries in the postwar world. But it is important to note that neither had clearly expressed concern over the *loss* of genetic diversity; both still were interested in the benefits of improving plant crop species and solving the world's hunger problems.

While the international sector seemed to have been slow to become aware of the potential dangers of genetic erosion, concerns within the American research community were beginning to find a voice. The title of the 1956 Brookhaven Symposium was Genetics in Plant Breeding, and several presenters noted Harlan and Martini's warning twenty years earlier about losing something "irreplaceable" should modern varieties replace traditional landraces. In 1959, the American Association for the Advancement of Science (AAAS) held a conference entitled Germ Plasm Resources. One contributor, Jack Harlan, son of H. V. Harlan, reiterated his father's opinion and added a renewed sense of urgency to the problem: "Unfortunately, the geographic centers of diversity upon which we have depended so much in the past for our many sources of germ plasm are in great danger of extinction. Modern agriculture and modern technology are spreading rapidly around the world. New uniform varieties from the experiment stations are replacing the old mixed populations that have grown, in some cases, since the Neolithic. The old centers of diversity are disappearing and time is running out faster than most of us realize. Adequate and thorough exploration must be made *now* before it is too late" (Harlan 1961: 16).

This was an almost wholly utilitarian way of valuing genetic diversity; one should not mistake the interest in preservation as sentimental nature conservation. As Henry A. Wallace wrote for the same conference, "Wild plants and animals are not as important as human beings, but their potentials are such that we should have the deepest concern about any form of life disappearing completely" (Wallace 1961: 85). Still, there was a sense that the world would lose part of its "heritage"—represented in both scientific and cultural value—if certain landraces and wild relatives were allowed to go extinct. In particular, primitive cultivars that revealed information to modern researchers about early human societies or about the evolutionary development of modern species were of special interest. In addition, there was always an interest in those overlooked species that might contain valuable germplasm for breeding purposes.

One interesting characteristic in the comments of those who were sounding the alarm was that the idea of setting aside acreage in the "centers of diversity" to preserve wild relatives "in situ" was seen as an unreasonable solution. A far more practical plan would be to collect specimens, give them a place in storage, and allow the planting of modern varieties to continue. As Jack Harlan noted, "A predicament now exists in which the technologically backward countries cannot afford to keep their great varietal resources and the more progressive countries cannot afford to let them be discarded. The only answer is an extensive exploration and collection program devoted to assembling as much of the germ plasm of the world as possible and diligent maintenance of the material once it is obtained" (Harlan 1961: 16). This solution, of course, was a guaranteed protection against the "backward" countries who would invariably—in the eyes of those in the "forward" countries—allow their germplasm to be lost. It was practical in the sense that one could hardly expect any developing nation to hold back from planting all available acreage with the most efficient crops available.

Otto Frankel and Protecting Genetic Resources

While the international organizations involved with agriculture did not seem particularly concerned with losses of genetic diversity, interest in conservation was sparked when a new international research program was proposed. In the early 1960s, the International Union of Biological Sciences (IUBS) began to promote the idea of an International Biological Programme (IBP), a long-range project that would examine the status of the earth's biological systems through various investigative committees made up of scientists from around the world (Blair 1977: 1–3). In 1964, the first draft program of the IBP recommended that "a survey be made of available knowledge of genetic variation of major crop plants and their wild relatives to determine areas of the globe where intensive exploration should be made for additional germplasm material" (quoted in Frankel and Hawkes 1975: 1). Although the year before, the FAO conference had recommended that a committee to be named the Panel of Experts on Plant Exploration and Introduction be established, the FAO had yet to select the members of the panel (Harlan 1975: 619). The IBP, in contrast, had formed a Gene Pools Committee and had selected as its chair Austrian-born Sir Otto Frankel, a noted plant geneticist and breeder who was residing in Australia. This move encouraged the director-general of FAO, Dr. B. R. Sen, to order a review of the FAO's progress in plant exploration and introduction. In an effort to coordinate with the IBP, Sen asked Frankel if he would be in charge of the review as the chair of the FAO Panel of Experts. Frankel agreed and the IBP/FAO partnership began; it would continue for the next ten years (Frankel and Hawkes 1975: 2).

The arrival of Frankel on the international scene may have been one of the key deciding factors for the increase in awareness in the 1960s of the consequences of the loss of genetic diversity. His personal concern for germplasm conservation was evident in conferences and publications of the late 1960s and early 1970s. But in addition, it was evident that scientists working in regions important for their genetic resources were beginning to report "a greatly accelerated rate of displacement of primitive crop varieties by locally selected or introduced cultivars" (Frankel and Hawkes 1975: 1). This situation was made apparent in the presentations of the 1967 Technical Conference on the Exploration, Utilization, and Conservation of Plant Genetic Resources, jointly sponsored by the FAO and IBP. Arguably, this was the first international convention that held as a central theme the need to *conserve* genetic resources. Statements of concern in previous years had been found in scattered conference papers; now the concern served as the main focus of the discussion. As the new FAO director-general A. H. Boerma wrote in the foreword to the conference proceedings: "The full use of the genetic resources present in the primitive crop races is just beginning. . . . The loss or destruction of the world's genetic resources by short-sighted or ill-directed planning is something so wasteful that its consequences can be disastrous both for world food production and for man himself" (Boerma 1970: viii). Although most of the papers presented and published were technical pieces on sam-

pling methods, exploration tactics, and taxonomic reviews, there was
an overall air of urgency that was absent from the previous FAO conferences.

Frankel turned out to be one of the most outspoken defenders of the
conservation of germplasm. He is even given credit for coining the
term *genetic resources* in its contemporary sense (Witt 1985: 123), although his co-editor for the 1967 conference publication, Erna Bennett, is said to be the first to have popularized the phrase *genetic conservation* (Fowler and Mooney 1990: 149). Their joint introductory
article to the 1967 conference volume (published in 1970) arguably
was the inaugural address that marked the launch of the modern
germplasm conservation movement. The authors warn that "the tendency towards a narrowing genetic base has intensified in recent years
as a result of the widespread introduction of cultural measures which
have done much to minimize or even remove environmental differences over wide areas" (Frankel and Bennett 1970: 8). Frankel and
Bennett, both world-renowned plant breeders, acknowledged the
wonderful improvements promulgated by modern breeding techniques, and they held up as models the new wheat and rice varieties
developed by the Rockefeller Foundation. "Yet their success," they
wrote, "represents a very real and immediate threat that the treasuries
of variation in the centres of genetic diversity will disappear without a
trace" (Frankel and Bennett 1970: 9). Echoing Harlan and Martini's
words in 1936, they conclude that "the extinction of the natural
sources of adaptation and productivity represented by primitive varieties may turn out to be an irreparable loss to future generations"
(Frankel and Bennett 1970: 12).

One of Frankel's own articles in the collection (he wrote or co-wrote
four sections of the edited volume) contained revealing comments
about the differences between "gene pool conservation" and traditional "nature conservation." "In principle we are here on common
ground with nature conservation in general; but there is one important
difference. Nature conservation aims to protect areas representing
habitats and communities which can be identified. Gene pool conservation goes further. It is concerned with genetic differences which often can only be surmised, but not identified. It is therefore concerned
with population samples, possibly along latitudinal or altitudinal tran-
sects, often over extensive areas; hence a 'genetic reserve' should in-

clude a spectrum of ecological variability so as to provide a spectrum of genetic variability. It may therefore have to be either extensive, or scattered—the latter, as conservationists know, being difficult to manage" (Frankel 1970a: 473). Frankel succinctly identified here the central challenge to conserving genetic variety in the field and likely the main reason why previous observers had dismissed conservation in the wild as impractical. Since varieties could be located in small pockets along environmental gradients, they might be difficult to protect because of their vulnerability to perturbation. But most important, Frankel was making a clear statement in support of preserving germplasm in its natural environment (in situ) as opposed to in a gene bank (ex situ). He certainly was not discounting the importance of collection and storage, but as he declared in his conclusion, "Conservation of wild plants is most effective in their natural environment. . . . Where necessary, protection in reserves should be sought, either in connection with, or as an extension of existing systems of conservation" (Frankel 1970a: 487). Frankel in this way was linking the concerns for genetic resources with other nature conservation concerns, including efforts to preserve species and ecosystem types. Although he made a clear distinction between protecting gene pools and preserving "nature," he also associated them closely in his final comments, likely realizing that such conservation causes would work better if linked together. Here perhaps was an early precursor to the inclusive quality of the biological diversity paradigm that would develop over the next decade.

The Southern Corn Leaf Blight

Even with all of the new concern about protecting genes in the wild, there were not many examples to show the developed world why it should worry about the loss of such resources. Breeders had produced crops with higher yields and with stronger resistance to the diseases that had plagued farmers for centuries. The Green Revolution had spread throughout the developing world with great success, and therefore there were less people suffering from malnutrition. For a country like the United States, which dominated the world markets in wheat, corn, and soybeans, the problem of losing genetic resources in foreign wildlands must have seemed like a low priority.

But in 1970, the American corn crop was hit with a fungal disease that became known as the southern corn leaf blight. The blight was responsible for a 50 percent drop in yields in some southern states; the total losses nationwide amounted to 15 percent (NRC 1972: iii). The sweeping impact of the blight, which in previous years had been identified in isolated patches, was largely the result of a certain kind of cytoplasm that had been used in developing the hybrid seed used by a majority of American farmers. Apparently, a new strain of fungus had adapted quite well to infecting this particular cytoplasm. The genetic uniformity of the corn fields, in combination with unusually warm, moist weather, allowed the fungus to spread rapidly across the southern portion of the corn belt, leaving many farmers in financial ruins. Fortunately, in later summer months, the weather became cooler and drier overall, and the fungus was not able to spread into the northern states as dramatically as it had in the South. In addition, the USDA was able to mobilize scientists to develop a solution to the problem, and corn yields were back to normal the very next year (Doyle 1986: 1–7). But the damage had left an indelible mark in the thinking of both the American and the international agricultural communities. The world had witnessed the mighty modern power of U.S. agriculture humbled by a lowly fungus. The genetic uniformity of the corn crop had revealed a major weakness, and developed countries now had a reason to be concerned with maintaining the world gene pools: potentially valuable germplasm needed to be available for use in emergency situations.

The corn blight prompted the National Research Council to undertake a survey of other U.S. crops. Published in 1972, *Genetic Vulnerability of Major Crops* revealed some disturbing numbers. The study showed that many crops were dominated by only a few major varieties. For example, 96 percent of the pea crop was planted with only two pea types, and 95 percent of the peanut crop consisted of only nine varieties of peanut (NRC 1972: 286–287). Other crops, especially the larger money makers, were not quite so uniform, but the overall conclusion of the report was that "most major crops are impressively uniform genetically and impressively vulnerable" (NRC 1972: 1). The study served to bring the issue of genetic vulnerability and the corollary concern of genetic erosion even further into the media spotlight.

International Efforts: The IBPGR and the 1972 United Nations Conference in Stockholm

In the meantime, advances were being made on the international front. This time, the U.S.-backed interests were proposing a global network of research centers that would lead in collecting and evaluating existing germplasm. With the success of the IRRI and CIMMYT, the Rockefeller and Ford Foundations in conjunction with the World Bank were looking to expand the number of International Agricultural Research Centers (IARCs) and establish an infrastructure to link them all together. By 1972, four other centers opened up, in Columbia (1967), Nigeria (1968), Peru (1971), and India (1972) (Kloppenburg 1988: 161). In many ways, this vision was in direct competition with the FAO/IBP efforts. According to some policy analysts, the ensuing struggle over the control of genetic resources research was a characteristic example of the developed countries (represented by the foundations and the World Bank) trying to protect their interests in the developing world (represented by the FAO and the UN). As Pistorius and van Wijk observe, "Both parties had different but strong reasons to take the lead. The FAO's mandate to create food security made conservation of land races a logical start for more research on food crops. The two Foundations and the World Bank, however, were specifically interested in strengthening the Green Revolution in areas susceptible to communism. By locating IARCs precisely in these areas, political turmoil could be prevented. Also the 'seed flow' between the IARC collections could be of use for the agro-industry in the USA and other industrialized countries" (Pistorius and van Wijk 1999: 96–97).

U.S. interests did not want the FAO to control the course of global conservation of genetic resources because that would mean equal voting power between developed and developing countries, and the numerous Third World nations would easily be able to advance their interests against the usually more powerful industrialized states. In 1971, in an attempt to attract the support of the developing world, the foundations and the World Bank established the Consultative Group on International Agricultural Research (CGIAR), an organization billed as an opportunity for donor countries and groups to meet with agricultural experts from developing countries to identify new regions that might be most appropriate for new research centers (Pistorius and van Wijk 1999: 98).

Certainly, with all of the work on genetic conservation that the FAO and the IBP had undertaken through the 1960s, it would seem that their network and resources would enable them to bring together research information more efficiently than the new CGIAR. The FAO had even created a Crop Ecology and Genetic Resources unit in 1968, which had served as a clearinghouse for information and had developed a computerized data bank cataloguing the genetic resources available to breeders in gene banks around the world (Witt 1985: 123). But the FAO stayed focused on facilitating information and seed exchange, whereas the CGIAR had already established centers devoted to research and collection. Ultimately, a compromise had to be reached, because while the FAO lacked the organized network, CGIAR and its big-money donors did not have the complete trust of the developing countries. The answer was to create a new "conservation institute" financed by the CGIAR but operating within the bureaucracy of the FAO. After several years of negotiations, the International Board of Plant Genetic Resources (IBPGR) was established in 1974. Those who supported the IBPGR hoped the new institute would enjoy the best of both worlds: funded by developed nations but protected under the UN umbrella, the organization was able "to set up an international seed exchange network that went far beyond the CGIAR seedbanks. In 1984, only ten years after its establishment, the IBPGR network bound together about 40 national collections in 30 countries" (Pistorius and van Wijk 1999: 99).

But some still believed that the IBPGR was just a front for the developed world to milk the developing world of its most valuable genetic resources. Instead of the humanitarian public image of the FAO, the IBPGR was perceived as having far more selfish foundations. As Kloppenburg notes, "Sixty-nine percent of the IBPGR's 1984 budget was underwritten by just six . . . donors: Canada, Japan, Netherlands, United Kingdom, United States Agency for International Development, and the World Bank. . . . The board's policies are set not by debate among member nations of the FAO but through decision-making processes internal to the CGIAR. The IBPGR may cloak itself in the 'internationalist' legitimacy provided by its association with the FAO, but the board is not subject to the control of the United Nations. The financial heart and political soul of the IBPGR lie elsewhere" (Kloppenburg 1988: 164). In short, the system was perceived as a contemporary version of colonial imperialism. "The CGIAR system is, in one sense, the modern successor to the eighteenth- and nineteenth-century botanical gardens that served as conduits for the transmission of plant genetic information from the colonies to the imperial powers" (Kloppenburg 1988: 161).

Others, however, view this particular chapter in the international movement for genetic conservation as a fine example of initiative, foresightedness, and global cooperation. Without the push from the foundations (whatever their motives), perhaps the actual protection of germplasm resources would have taken longer to be operationalized. As Wilkes writes, "The issue of genetic resources is a dramatic model of the importance of consensus, infra structure (UNEP, IBP, CGIAR) and international will from the nonscientific public policy sector. The scientific recognition of the problem and the methodologies were in place by 1968 but the institutional arrangements and support were not fully functional until later. Without the surge of environmental awareness and the role of key individuals in the 70s the international issue of genetic resources might still be floundering" (Wilkes 1983: 162). Whatever opinion one has of the political positioning that went on, it is evident that genetic resources had become an issue of paramount importance to powerful countries and organizations.

The clearest statement of the newly recognized value of genetic diversity, however, came out of the 1972 United Nations Conference on the Human Environment, in Stockholm. The conference recommendations 39–45 dealt with the concern for genetic resources, and although most of the proposals for action were modest, there were still clear expressions about the importance of genetic diversity. Recommendation 40, for example, begins with these words: "It is recommended that the Governments . . . make inventories of the genetic resources most endangered by depletion or extinction: (a) all species threatened by man's development should be included in such inventories; (b) special attention should be given to locating in this field those areas of natural genetic diversity that are disappearing" (in P. Stone 1973: 165).

Because of the UN sponsorship of the conference, Otto Frankel (as the FAO-designated expert on genetic resources) exerted significant influence in bringing these recommendations to the table. The recommendations (and Frankel's celebrity) helped to ensure that the issue of genetic conservation would share the media spotlight of the conference in general (Fowler and Mooney 1990: 150). Unfortunately, it seems that individual governments did not feel that it was urgent to take action on many of the Stockholm recommendations, particularly those tasks that were being handled by other interests. For example, the survey of genetic resources suggested in recommendation 40 was in fact already underway through an FAO effort led by Frankel himself. But while governments may not have taken direct action, the conference still was considered successful, not only for raising important issues and bringing them to popular attention but also for inspiring other interested parties—such as numerous nongovernmental organizations in various countries—to mobilize and work for solutions to pressing environmental problems.

Advances at the Molecular Level: Drugs and Recombinant DNA

One other related topic that enjoyed some popularity in the literature in the 1960s and 1970s were the remarkable medicines that had been discovered recently in wild plants. In a time when most people believed that science was manufacturing all of the potent drugs that were extending people's lives, the fact that certain important pharmaceuticals were derived directly from plants had a particular attraction. A book by Norman Taylor, *Plant Drugs That Changed the World* (1966), told the stories of several drugs derived from the plant kingdom: cocaine, digitalis, quinine, resperine, morphine, steroids. Taylor purposely chose to write in a nontechnical style, focusing more on making fables out of the facts and trying to draw attention to the marvels of nature rather than human technical innovations.

An edited volume entitled *Plants in the Development of Modern Medicine* (1972) was based on papers given at a 1968 symposium at Harvard. The introduction notes that "today there is probably more interest in drugs derived from plants than at any other time in history" (Mangelsdorf 1972: xi). As one contributor to this collection put it, the discovery in the 1950s and 1960s "of a series of so-called 'Wonder Drugs,' nearly all from vegetal sources, sparked a revolution. It crystallized the realization that the Plant Kingdom represents a virtually untapped reservoir of new chemical compounds, many extraordinarily biodynamic, some providing novel bases on which the synthetic chemist may build even more interesting structures" (Schultes 1972: 104). It was an exciting time, for not only were new discoveries being made but techniques to refine compounds had also advanced dramatically, opening up many new opportunities for scientists the way that new breeding techniques had created opportunities for geneticists in the first decades of the century. In addition, it was considered a wide open field; "even the flurry of phytochemical research engendered by the 'Wonder Drugs' has barely scratched the surface of the Plant Kingdom" (Schultes 1972: 105). The possibilities of what might be found in the wild and employed for human health seemed limitless.

Then, in 1973, a scientific breakthrough of major proportions took to a new level the application of genetics to medical and agricultural research. Stanley Cohen of Stanford University and Herbert Boyer of University of California in San Francisco cut a section of bacterial DNA in half, inserted a foreign gene, and recombined the ends on either sides of the new gene. It was the first example of "recombinant DNA," although its more popular title came to be "genetic engineering" (Witt 1985: 123). While it would take several years before techniques were refined such that they could be put to any practical use, the implications of the success of Cohen and Boyer's experiment were enormous. At the turn of the century, the discovery of the gene had expanded the interest in ancient landraces and wild relatives of crop species because breeders could now use "characteristics" of plants once dismissed as inferior varieties. With the advent of biotechnology, the entire gene pool of all living things carried potential value. Theoretically, geneticists could now extract a gene from any one organism, plant or animal, and splice it into the DNA of another living thing. The results, of course, might not always turn out as expected or intended, but the technology was now available and thus the possibilities truly were endless. However, because most of the early research in genetic engineering was in the world of medicine, these new recombinant techniques did not significantly impact agriculture until the 1980s. But the promise of biotechnology, not only for agriculture but for any interest that used living resources, led to a widespread support for preserving biological diversity as a potential source of useful genetic material.

*The 1970s: Expanding the International
and Domestic Infrastructures*

While applied biotechnology was still some years off, the interest for protecting germplasm for traditional breeding purposes grew stronger every year throughout the 1970s, spurred by the activity in the early part of the decade that had established the infrastructure for international genetic conservation. A second FAO/IBP conference on germplasm resources, entitled the Technical Conference on Plant Exploration and Introduction, took place in Rome in 1973. Using papers from this conference, Otto Frankel and J. G. Hawkes edited the 1975 publication *Crop Genetic Resources for Today and Tomorrow.* This book served as the authoritative text on FAO/IBP efforts to gather information on existing germplasm resources, as well as on exploration, collection, and conservation techniques. The editors' personal analysis of how far the perception of genetic diversity had come is worth highlighting. In their introductory article, "Genetic Resources: The Past Ten Years and the Next," Frankel and Hawkes wrote: "Perhaps the biggest transformation of the last decade has been the development of widespread awareness and concern. Ten years ago, as we said at the beginning of this chapter, the problem of 'genetic resources' was unknown. In the last few years, it has sunk into the consciousness of scientists and administrators and of the many people who have become concerned about the resources of the earth" (Frankel and Hawkes 1975: 4). They noted the recommendation made in Stockholm in 1972 and the collective work of the FAO, IBP, CGIAR, and IBPGR. With the network of IARCs in place, the basic framework was set for saving "the world's fast diminishing resources." The volume overall carried a hopeful tone. While there was much work to be done, Frankel and Hawkes conveyed a sense that the concern, knowledge, will, and support were all available to complete the tasks at hand.

In the United States, a national infrastructure was likewise being constructed. The USDA took steps to expand the National Plant Germplasm System, forming their own advisory board according to the model established by the FAO and CGIAR. In addition, more U.S. organizations began to show an interest in the issue. The 1975 American Agronomy Society meeting focused on the lack of a national policy for germplasm protection. A 1976 symposium sponsored by the Amer-

ican Association for the Advancement of Science was entitled Plant
Germplasm Resources: American Independence Past and Future
(Wilkes 1983: 150). In the same year, the Society for Economic Botany
held a conference that focused on the prospects of improving crops,
the possibilities of new food resources, and the industrial markets for
nonfood crops based on wild plants presently underutilized by the
Western world (Seigler 1977). Overall, concern for genetic resources
had reached unprecedented heights. Now, instead of there being only
a few interested organizations, the urgency to protect germplasm had
spread throughout many different constituents. The issue of maintaining genetic diversity was developing a wide base of support.

This concern, as expressed in the United States, was perhaps best
encapsulated in this period by the 1978 National Research Council report *Conservation of Germplasm Resources: An Imperative.* This publication, which had an urgent tone, presented genetic diversity as the definitive yardstick by which to measure all conservation efforts: "Saving
the rich diversity of genetic material that has been provided by natural
mutation and evolution can be achieved and is worth whatever effort
may be required. It is critically important that the people of the United
States recognize the long-term dangers inherent in the loss of specific
genes and of genetic diversity, recognize that diversity in germplasm is
an essential national resource, and treat it as such" (NRC 1978: 4). Saving species and ecosystems, in the eyes of the authors, was a strategy for
preserving the global gene pool. Because the gene was the common de-
nominator of value, if one maintained diversity at the genetic level one
could likely guarantee the protection of both species and ecosystems.
Also significant in the 1978 NRC report is the early use of the term *biological diversity,* in conjunction with the term *germplasm resources,* before the publication of the first definition of biological diversity in the
1980 CEQ *Annual Report.* It is evident that the two terms are not considered to be synonymous; instead, it seems that the authors thought of
biological diversity as the entire structure of component parts of the
living biota. But the measure of a healthy ecosystem, or even a healthy
species, was whether the genetic diversity of the ecosystem or species
was fully intact. Biological diversity was never defined in the report,
and while it was used only a few times, it was obvious that when the authors employed the term they were trying to link the conservation of
genetic diversity to a larger perspective.

Revisiting the 1981 Strategy Conference

In concluding this discussion on the agricultural interest in genetic diversity, it is worthwhile to take another look at the 1981 Strategy Conference on Biological Diversity and the interests that were represented. The principal conclusions of the conference were largely related to concerns for germplasm and genetic diversity. This fact likely had some connection to the sponsors and invited conferees. With the Department of State and the Agency for International Development taking the lead, an observer would likely conclude that the conference was intended to examine the relationship between conservation and economic and national security issues, and indeed, the main agricultural theme was characterized by commercial and international concerns. The list of speakers included Dr. William L. Brown, chairman of Pioneer Hi-Bred International, whose introductory presentation was entitled Genetic Diversity: Serious Business for Crop Protection and Maintenance, and Dr. J. T. Williams, executive secretary for the International Board for Plant Genetic Resources. But in addition to the focus on crop species and germplasm for agriculture, there were also statements linking the concern for genetic diversity to concerns for species and ecosystems. While it seems that the 1981 conference was largely inspired by the germplasm resource crisis, the inclusion of discussions involving species and ecosystem conservation was a significant step historically in tying the different hierarchical levels together under one conservation concept.

Certainly, the concern for plant genetic diversity—of crop species in particular—remains a cause with its own momentum. Books, papers, and conferences are still produced on the topic of germplasm resources, but its place within the broader concept of biological diversity has been beneficial for both causes. It is evident that in the earlier years of the 1980s, the concept of biological diversity received a huge boost from the community that had been fighting to protect genetic diversity many years before the term biological diversity had fully established itself in the conservation vocabulary.

Conservation Biology
The Roots of a "New" Discipline

The third group interested in genetic diversity, the group that most consciously tied the concern of genetic diversity to related concerns for species and ecosystems, was the community of scientists who came to be known as conservation biologists. The discipline of conservation biology seems to have co-evolved with the concept of biological diversity. Both were created more or less as part of the reaction to the extinction "crisis" that was gaining recognition in the 1960s and 1970s, and both appealed to a wide interest base, drawing adherents to the cause of conservation from many different disciplines. Indeed, when the Society for Conservation Biology was first established in 1986, it was described as "a response by professionals . . . to the biological diversity crisis" (Soulé 1987: 4). In essence, to call oneself a conservation biologist is the equivalent of declaring one's concern for the maintenance of biological diversity.

The origins of conservation biology lie in the 1978 International Conference on Conservation Biology. The book containing papers from the conference, *Conservation Biology: An Evolutionary-Ecological Perspective,* edited by Michael Soulé and Bruce Wilcox, was published two years later. Soulé has long been considered as the founder and zealous leader of the conservation biology movement, and his early writings are seminal works for the field. Soulé is a geneticist by training, concerned with the genetic fitness of wild populations of plants and animals. From the beginning, then, conservation biology contained a significant component of genetic studies and relied on research done in the field of population biology, which itself was dependent on a clear understanding of how genetic variation impacted the structure and behavior of populations. Conservation biology was never limited to the genetic perspective. As Soulé and Wilcox wrote in 1980, "With regard to breadth, conservation biology is as broad as biology itself. It focuses on the knowledge and tools of all biological disciplines, from molecular biology to population biology, on one issue—nature conservation" (Soulé and Wilcox 1980b: 1).

But it is evident that Soulé and Wilcox felt that it was time to bring genetic research of wild populations out of the sterile laboratory and into the field. "Pure" science, claimed the authors, had for too long held the "applied" sciences in disdain. With the warnings of an impending "extinction wave," especially in tropical habitats, Soulé and Wilcox declared that "the luxury of prejudice against applied science is unaffordable" (Soulé and Wilcox 1980b: 2). The authors then singled out "conservation genetics as an example of such academic disinterest," proclaiming that "one of the purposes of this book is to begin correcting this handicap" (Soulé and Wilcox 1980b: 2).

The next definitive publication for conservation biology was *Conservation and Evolution* (1981), by Frankel and Soulé. This book more clearly established the concern for genetic diversity as a key plank in the platform of the new discipline: "In this book we focus on the plight of rare species, existing or potential—species whose effective population sizes are small. The reason for this focus is simply that the genetics of nature conservation is the *genetics of scarcity.* That is, our concern is with fitness and evolutionary potential concomitant with loss of genetic variation" (Frankel and Soulé 1981: 8). With this publication, it was apparent that Frankel wanted to report the losses in genetic diversity not only to the agricultural and plant-breeding community but also to fellow scientists in different fields. Frankel's work, as we have seen, had largely concentrated on crop species and their wild relatives. By teaming with Soulé and expanding the genetics discussion to wild populations he was reaching a whole new audience. Additionally, he was a direct link between the rise of conservation biology and the longer history of concern for germplasm resources, especially those of wild relatives of domesticated species.

Conservation and Evolution covered all of the basic theory and methods for managing populations of plants and animals in accordance with their genetic health and vigor. "In this book," the authors wrote, "we attempt to bring together the genetic principles for the conservation of all forms of life, wild or domesticated, lions or lizards, oaks or orchids, cattle or ducks, rice or potatoes. The unifying factor underlying survival and adaptation, in time and in space, is genetic diversity; and the nature, distribution and preservation of genetic diversity is the central theme of this book" (Frankel and Soulé 1981: vii). Frankel and Soulé did not distinguish whether wild or domesticated gene pools were more important; rather, they were chiefly interested in protecting evolutionary processes. They emphasized this characteristic in an explanation of their choice of the term *conservation* in their book title. In this explanation, they drew a distinction between "cultivated" germplasm and "wild" gene pools: "Genetic resources of domesticates are *preserved* not for their own sake, but because of their immediate or potential usefulness to man, be it in breeding or in some form of research. The reason for nature *conservation,* as we see it, is diametrically different. Its essence is for some forms of life to remain in existence in their natural state, to continue to evolve as have their ancestors before them throughout evolutionary time" (Frankel and Soulé 1981: 6).

For conservation biologists, maintenance of genetic diversity was a call for the protection of evolutionary dynamism. It was not an argument simply based on the humanly beneficial products of genetic variability, although Frankel and Soulé did not ignore this element of conservation. The goal of providing conditions for the continuation of life and life processes was the central mission of conservation biology. Not only did this articulated objective mark a significant enlargement over the more economic concerns for genetic diversity, it also signaled a shift in the thinking about methods for maintaining wild animal and plant populations. Traditional wildlife management had long valued species as game or for some other recreational purpose. The concern for endangered species had wrought some changes in management techniques, but now conservation biology was raising the stakes even higher. Society ought to be concerned with "all forms of life," as Frankel and Soulé wrote, and wildlife managers would need new methods and information if we wish to maintain "some forms of life . . . in their natural state." This objective was one of the central goals of those who were to write about the loss of biological diversity. Frankel and Soulé were approaching and framing the issue from the molecular viewpoint of genetic conservation.

Population Biology and Island Biogeography

As mentioned by Soulé and Wilcox, another significant parent discipline to conservation biology was a field of study called population biology. The research in population biology mostly focused on the ways in which populations—their numbers, structures, functions—fluctuated over time, and the reasons why such fluctuations occurred. The behavior of a population over time obviously was dependent upon its genetic structure and the interaction of its inherent genetic variation with the surrounding geophysical environment. The interest of conservation biologists in population biology is evident—the better one understands how a population functions and changes over time, the easier it is to protect that population from extinction. One of the leaders in this field was Robert MacArthur, who in 1966 published a textbook with Joseph Connell, *The Biology of Populations*. Also during this period, MacArthur was developing, in partnership with E. O. Wilson, the theory of island biogeography, which was an attempt to explain the distribution of populations, their fluctuations in numbers, and ultimately the relationship between the total area of an ecosystem and the diversity contained therein. Biogeography itself had a venerable history, but it had mostly been a descriptive science—collecting information about "the distributions of species and higher taxa and the taxonomic compositions of biotas." MacArthur and Wilson were ready to go one step further: "Biogeography appears to us to have developed to the extent that it can be reformulated in terms of the first principles of population ecology and genetics" (MacArthur and Wilson 1967: 181–183).

The theory was essentially based on studying the flora and fauna of islands, noting how species variously established themselves, flourished, or became extinct. The first chapters of the 1967 book *Theory of Island Biogeography* were dedicated to considering area-diversity curves, the mathematical relationship between the amount of island land available for terrestrial organisms to populate and the total diversity that the land was capable of supporting. This research was seminal for the projections of species extinctions that would appear in the late 1970s, for although researchers (especially conservation biologists) concerned with species loss were focused mostly on the tropical parts of continents (not small islands) the theory of island biogeography still had significant meaning for the fate of species in fragmented habitats. Patches of ecosystems broken up by development had become like islands for plant and animal inhabitants; such populations were isolated from one another. MacArthur and Wilson noted this application in their first chapter: "The same principles [of island biogeography] apply, and will apply to an accelerating extent in the future, to formerly continuous natural habitats now being broken up by the encroachment of civilization" (MacArthur and Wilson 1967: 4).

The implications of this new theory set off a contentious debate among conservation biologists in the mid-1970s about the size and shape of nature reserves. When Daniel Simberloff and Lawrence

Abele (1976), using the area-diversity relationship developed by MacArthur and Wilson, suggested that numerous small reserves might conserve a more diverse array of species than one large reserve, they drew a quick and testy response from other scientists working on conservation problems (see Diamond 1976; Terborgh 1976; Whitcomb et al. 1976). The issue became known as the SLOSS (Single Large or Several Small) debate, and it revealed the fundamental tension between those scientists who wished to answer questions about population biology objectively and those who were personally motivated to protect as much of the natural world as possible.

While the SLOSS debate did not directly involve concerns for maintaining the genetic fitness of species, the prominence of population biology and the need for reserves represented the other key characteristic of conservation biology related to worries over the maintenance of genetic diversity. As mentioned, conservation biologists were extremely concerned about maintaining an environment where evolutionary and ecological processes could function unhindered. As Frankel and Soulé wrote, "All that is attempted is to provide conditions, based on our best scientific insight . . . which will make it possible for an evolutionary succession of organisms to continue" (Frankel and Soulé 1981: 7). Here, the central concern for genetic diversity was placed within the larger context of process. Conservation biologists believed that without genetic variability, evolution would be deprived of its basic raw material. Alternately, without the conditions conducive for evolution to take place, genetic diversity would lose its primary significance. The recognition of this particular ecological value characterized the period in which both the discipline of conservation biology and the concept of biological diversity came into existence. The value of genetic diversity to the maintenance of all life on earth was first recognized by those who were trying to create a unified picture. Conservation biology provided the lens for those focused solely on the genetic level to see how genetic diversity could be perceived as the essential foundation for the conservation of the natural world and all of its ecological processes and components.

Precursors to Conservation Biology

While the advances in population biology and applied biogeography provided some of the foundational structure for conservation biology, there were several publications that represented more direct precursors. Soulé and Wilcox list several sources of the new "mission-oriented discipline comprising both pure and applied science" (Soulé and Wilcox 1980b: 1), referring the interested reader to Dasmann's *Environmental Conservation* (1968) and Ehrenfeld's *Biological Conservation* (1970). Dasmann's text (which was first published in 1959) was a general textbook on environmental problems and included lengthy sections on traditional natural resource issues, including agriculture, water, timber, fisheries, recreation, and the growth of the human population. Curiously, there was only a short chapter on "The Management of Wildlife"; although Dasmann's tone in this chapter revealed a link to conservation biology, in general his subject matter stayed close to the typical concerns of traditional wildlife management. One example of Dasmann's enlightened perspective, however, appeared in his concluding paragraph to the wildlife chapter: "Where wildlife still remains, it is essential to take early steps toward conservation and management; with present rates of human population increase it is unlikely that any area will remain unaffected by man for long. It would be unfortunate if future generations were denied first-hand knowledge of the variety of wild animals which have accompanied and influenced man throughout his evolution and spread over the earth" (Dasmann 1968: 253). The text was certainly ahead of its time (especially the first edition), and it likely influenced many future conservation scientists with its all-inclusive treatment of environmental issues and concern for the diminishing quality of the natural world.

Ehrenfeld's *Biological Conservation* (1970) represented a more important antecedent, and the similarity of the book title to the name of the future discipline was likely not by chance. Ehrenfeld stated in his preface that, in his opinion, "a comprehensive sample of existing animal and plant species and natural communities should be preserved" (Ehrenfeld 1970: 2). Although no single section of the book specifically addressed the loss of genetic diversity, Ehrenfeld often referred to the problem of disappearing species in genetic terms. For example, in order to make decisions about preserving species, Ehrenfeld declared, "it is necessary to know what fraction of the total gene pool is threatened" (Ehrenfeld 1970: 187). In another chapter, "Factors That Threaten Species," Ehrenfeld used the example of how the coyote, by breeding into the Texas wolf population, was threatening to wipe out the genetic uniqueness of the former species through hybridization (Ehrenfeld 1970: 95). Ehrenfeld also emphasized the importance of protecting the engines of evolution, not simply the genetic diversity of the flora and fauna but the conditions which allow "that nothing in the existing natural order [be] permitted to become permanently lost as the result of man's activities" (Ehrenfeld 1970: 4). But perhaps more significant was Ehrenfeld's plea to his fellow scientists to break out of their laboratories to address the world's environmental problems. Foreshadowing the arrival of conservation biology, he wrote, "Qualified biologists are beginning to forge a discipline in that turbulent and vital area where biology meets the social sciences and the humanities. The need is now very great for a scientifically valid presentation of the biological problems that are most relevant to the life of modern man" (Ehrenfeld 1970: vii). Ehrenfeld, who in 1987 would become the first editor of the journal *Conservation Biology,* was one of the more influential thinkers laying the foundation for the new discipline.

In addition to Ehrenfeld's book, a British periodical of the same name had begun publication in 1969. *Biological Conservation* was touted on its title page as "the International Quarterly Journal devoted to scientific protection of plant and animal wildlife and all Nature throughout the world, and to the Conservation or rational use of biotic and allied Resources." The journal published a wide range of articles, from examinations of species life histories or the ecological characteristics of a particular region, to descriptions of conservation programs in different countries, to discussions of the impact of human activities on ecosystems. An example of one influential article published in the periodical is Frankel's "Genetic Conservation of Plants Useful to Man" (1970). While focusing mainly on primitive cultivars, Frankel also mentioned the significance of the "protection in-situ of threatened and important wild communities" (Frankel 1970: 162). In the years following Frankel's article, *Biological Conservation* would publish more articles that discussed both the importance of maintaining the global gene pool and the specific genetic problems of certain populations of endangered species.

Another text worth noting as a precursor to conservation biology is a 1978 publication by Clarence Schoenfeld and John Hendee, entitled *Wildlife Management in Wilderness.* Among other topics, the authors discussed the "significant, though less obvious . . . role of wilderness as a hidden trove of those recessive genes necessary for genetic adaptability in the face of environmental change" (Schoenfeld and Hendee 1978: 26). These genes were important not only for the adaptability of the species, representing a "stock of variability that can be brought into play, sometimes within a single generation," but also for their potential usefulness to humans. While the authors generally used traditional methods for examining wildlife populations, their consideration of the genetic implications illustrates the trend in thinking about wildlife as a source of genetic diversity and the necessity of conserving that natural variability.

Finally, conservation biology's ideas about the maintenance of genetic diversity and its connections to protecting species and ecosystems also influenced publications in the 1980s. For example, the first seminal publication that saw wildlife management in terms of protecting genetic diversity was a 1983 volume, *Genetics and Conservation: A Reference for Managing Wild Animal and Plant Populations.* Edited by Christine Schonewald-Cox and several others, this volume rivaled Soulé's early publications as the most significant early work in the field. The ever-present Frankel provided the foreword, which opens with a brief historical summary: "Recently, population biology became an integral part of what is now termed conservation biology (Soulé and Wilcox 1980a). When it was postulated that genetic diversity may be a condition of long-term survival (Frankel 1970a; Frankel and Soulé 1981), population and evolutionary genetics were seen as having a significant role in conservation biology" (Frankel 1983: xiii). While Frankel might have been patting himself on the back as the original defender of genetic diversity, he does point out that the science of genetics and the concern for genetic diversity played an integral part in the formulation of the new discipline.

The editors of *Genetics and Conservation,* although eschewing the use of the term *conservation biology,* stated their aims in the preface, echoing the objectives laid out by Soulé in earlier publications: "Our goal in *Genetics and Conservation* is to contribute to the preservation of the natural diversity of species with a view to preserve the evolutionary potential of species by exploring the relationship between today's advances in genetics and their potential contribution to the quality of wildlife (animal and plant) conservation" (Schonewald-Cox et al. 1983: xvii–xviii). Indeed, the book was a comprehensive treatment of the concerns about genetic diversity in relation to the survival of species. Separate parts were dedicated to such subjects as isolation, extinction, founding and bottlenecks, hybridization and merging populations, and taxonomic considerations. It was the first volume, but certainly not the last, to focus the knowledge of genetics on the problems of protecting wild species.

When conservation biologists reached critical mass in 1986, the Society for Conservation Biology was founded. Planning for the new organization had taken place at a meeting held during the 1986 National Forum on Biodiversity. Since this time, conservation biology has become a popular subject for activist-minded students interested in the science behind conservation efforts, and many colleges and universities, especially those with environmental science departments, offer entire courses of study in the new discipline. The growing profession has also taken advantage of the many technological improvements that have been developed for research methods. Genetic techniques for managing species made huge strides forward in the 1980s and 1990s, as genetics research overall began to progress at an unprecedented rate. Improvements in molecular analysis, for example, have helped conservation biologists to measure the genetic variation in shrinking populations of endangered species, form theories about metapopulation structures and the patterns of genetic exchange, and manage the impact of nonnative species that hybridize with endemic populations (Smith and Wayne 1996b: v). Such research tools have helped conservation biologists answer the questions that first inspired the discipline, questions that are central to understanding the biological mechanisms of survival. The initial recognition that genetic studies were essential for understanding and predicting the population trajectories of endangered species and the role of genetic fitness in evolutionary and ecological processes is what gave the concern for genetic diversity such a significant role in the development of conservation biology.

Conclusion

The study of genes, from the earliest research in the first decades of the twentieth century to the present day, permeates the biological sciences, and the discoveries made since 1900 epitomize the sense of magic that scientific exploration can conjure. Both the scientific community and the general public view the gene as carrying the secrets of evolution, and thus its scientific value—as the source of the story of life on earth—is incalculable. In addition, anyone with knowledge of the science of genetics undeniably enjoys a certain amount of respect in Western society. The inclusion of the genetic perspective in conservation efforts has been an important step, not only for increasing the scientific tools and information available to conservationists but also for raising the work of conservationists to a higher level in the eyes of other scientists, politicians, economists, and everyday citizens.

The maintenance of genetic diversity has a practical component which is difficult to ignore. Genetic diversity's first line of defense has obviously been its utilitarian value, largely because the benefits that genetic diversity provides have touched a wide array of interests. Diversity's value as a source of breeding characteristics that agronomists could use to improve crop yields was recognized well before the discovery of the gene. In the first half of the twentieth century, the economic value of genetic diversity to agricultural interests was considerable; anyone who could produce superior crops had obvious advantages. The federal government's interest in genetic diversity illustrated the value to national well-being. In addition to utilitarian value, those interested in genetically unique "landraces" of ancient civilizations identified historical and cultural values in the preservation of these early crop species. Finally, with new advances in biotechnology, the full range of genetic variation could be employed to produce naturally based products of enhanced value. Thus, it was not simply specific genes that needed to be protected. The entire genetic code of life held possibilities of benefits for humanity, and this realization sparked the broad interest in gene conservation efforts.

To those within more traditional environmental circles, maintaining genetic variation provided support for other significant benefits, such as protecting species against extinction and setting aside large reserves to protect gene pools. Conservation biologists, who were trying to bring together the different hierarchical levels in the same way the concept of biological diversity would do, were motivated primarily by the plight of endangered species and the degradation of ecological processes and evolutionary potential. By using the tools of genetic research, conservation biologists were able to understand the details of specific conservation problems in an in-depth way and find solutions that helped to protect the ecological, aesthetic, and ethical values that inspired their work.

While not as broadly based or as popular as the concern for species diversity, the interest in protecting genetic diversity had a scientific pedigree and an urgency that could not be ignored. Because of its technical nature, it was a cause with which environmentalists were unlikely to be familiar. But the potential of values that the protection of genetic diversity represented, in combination with the powerful interests that recognized and defended these values, raised the issue into the upper echelon of conservation objectives. This position ultimately earned genetic diversity a place among the triumvirate of concerns that would come to define the concept of biological diversity.

5 *The Concern for Ecosystem Diversity*
From Parks and Natural Areas to
Biosphere Reserves

While the history of protecting ecosystem diversity presents its own unique set of issues, it compares in interesting ways to the history of concern for species and genetic diversity. Like early efforts to protect particular species, specific tracts of land were recognized as valuable for various reasons and were subsequently protected. For example, the idea of a preserve for maintaining a population of game species goes as far back in Western society as medieval times, when European nobility demarcated protected areas to serve as their hunting grounds (Thomas 1983). In the New World, land protection by the state and local governments was sometimes incorporated into town planning, where a common property would be reserved for use by all citizens, primarily for grazing livestock. Such examples of set-aside lands are often cited as early precursors to the more organized protection efforts of the late nineteenth and early twentieth centuries, such as parks, forests, and wildlife refuges. But while species protection efforts (through the species concept) clearly defined the kind of animal to be protected, there is some question as to whether all land protection efforts constituted attempts at protecting *ecosystems.* More often, scenic features or an abundance of resources, not ecological characteristics or the perspective of a natural area as a system, motivated the protection of land areas. Concern for ecological factors and for protecting a diversity of ecosystems came after initial attempts at protecting land in parks and preserves.

In fact, the ecosystem concept was not proposed until the mid-1930s and did not become a principal tool in the science of ecology un-

til after World War II. Like early definitions of the gene, the first definition of an ecosystem was somewhat malleable because scientists interested in ecosystems were not sure what an ecosystem looked like. In
addition, as in early genetics, competing concepts (the most popular
was the community concept) were used in ecological investigations.
But a significant part of the usefulness of the ecosystem concept came
in the flexibility of what defined an ecosystem. Originally conceived
solely as an object for scientific study, the ecosystem's boundaries
could be defined according to the need of the ecologist who was studying a particular part or characteristic of the natural world. But as a unit
of conservation, the ecosystem did not become part of environmentalism's popular vocabulary until the late 1950s and early 1960s, when
ecology was gaining wider attention. Although early efforts to protect
land were undertaken without the ecological knowledge of later conservationists, the success of such efforts to set aside large tracts of land,
and hence complete ecosystems, are worthy predecessors to more recent ecologically informed protection.

The evolution of land protection has had its own progression of values, which, as in the evolution of efforts to conserve species and genes,
have gradually broadened into the array of values that characterizes efforts to preserve biological diversity. For example, the national parks
movement is seen by many Americans as an effort to protect the grand
scenic features of the United States, thus emphasizing aesthetic, cultural, and recreational values. A closer look at the history of the parks
movement, however, also reveals that the early popularizers of the national parks idea recognized the economic value of scenic areas. In addition, the federal government's designation of national forests starting
about 1900 was a way to protect the utilitarian value of natural resources, and set-asides for wildlife refuges affirmed the close connection between the preservation of habitat and the maintenance of the
recreational, aesthetic, and scientific values that intact wildlife populations provided.

As the federal protection systems expanded in the early decades of
the twentieth century, the scientific community began to show an interest in protecting "natural areas" for their scientific value as outdoor
laboratories. The professional natural resource management community echoed the concern of the scientific community, expressing the
need to preserve intact representative examples of certain lands as

places to observe how humans have impacted the natural world. Not only did these concerns diverge in quality from the scenic and economic emphasis of earlier land protection efforts, but they also represented the first attempts to preserve a *diversity* of land types. Protecting *ecosystems* did not really become an environmental issue until the 1960s, when several international efforts inspired the scientific and conservation communities to look for ways to preserve an array of different ecosystem types all over the globe. It was also at this time that the international and national drive to protect ecosystems became inextricably linked with protecting species, as later it would be linked with protecting genetic diversity. A detailed look at the international efforts is warranted, for they heavily influenced the work of American ecologists and conservationists. By the 1970s, the ecosystem had become one of the most popular and durable units for conservation strategy throughout the Western world.

Most notably, maintaining a diversity of ecosystems gradually became synonymous with protecting the entire range of values that the natural world provided. Because of the inclusiveness of ecosystems, the benefits associated with ecosystem protection encompassed all of the values associated not just with the conservation of land but with the conservation of species and genes as well. These included utilitarian products, ecological services, scientific subjects, aesthetic resources, recreational areas, and cultural identification. Ecosystem protection has been seen as a way to provide homes for species and preserves for the global gene pool; for this reason, some view it as the most logical way to protect all three levels of biodiversity. Without ecosystems, one might ask, how can genes and species survive? But the same observation can be turned around: if species and genes were lost, ecosystems would lose components that would not only change their character but might also severely hinder their functional capabilities. In this way, the theme of interdependence is illustrated most clearly at the level of the ecosystem, in which each living part has a role to play.

The Scientific Foundation: Tansley's Concept of the Ecosystem

The concept of the ecosystem was first proposed by Arthur G. Tansley, a British ecologist, in his 1935 article "The Use and Abuse of Vegetational Concepts and Terms," published in the journal *Ecology*. The title of the article gives some hint as to Tansley's purposes in writing his piece, which largely was a response to a series of articles written by South African ecologist John Phillips. Phillips was, in Tansley's own words, the "chief apostle" to the organicist "creed" espoused in the popular and very influential theories of Frederic Clements (Tansley 1935: 285). In essence, Phillips was a proponent of viewing both plants and animals as members of a single biotic community whose processes and life cycles could be studied as if they were components of a single complex organism. Indeed, both Phillips and Clements went so far as to proclaim that biotic communities *are* organisms—living entities that operate at levels beyond present human understanding. Ecology was to be the holistic science that would study nature with this unifying organismal perspective.

Phillips unfortunately took a few too many liberties in his articles, one of which was to implicate the well-respected Tansley as a supporter of the organicist movement (Golley 1993: 13). Tansley, who in previous articles had voiced respect for the contributions Clements had made to ecology (particularly his work on succession), decided to take eloquent but pointed exception to Phillips's assertions. As Tansley declared, "At the outset let me express my conviction that Dr. Clements has given us a theory of vegetation which has formed an indispensable foundation for the most fruitful modern work. With some parts of that theory and its expression, however, I have never agreed, and when it is pushed to its logical limit and perhaps beyond, as by Professor Phillips, the revolt becomes irrepressible" (Tansley 1935: 285). Tansley goes on to critique Phillips's "abuse" of the concepts of *community* and *organism*. Tansley believed that animals and plants were "too different in nature to be considered part of the same community" and that the progressive step of calling such a community an organism was "objectionable because the term is already in common use for an individual higher animal or plant" (Tansley 1935: 306). Tansley did support one Clementsian term—*biome*—which referred to "the whole complex of organisms inhabiting a given region." He called the concept "unobjectionable, and for some purposes convenient" (Tansley 1935: 299). But for the most part Tansley found little use for the concepts and terms proposed by Phillips, and he particularly found distasteful the heavy organicist flavor of the new definitions of older ecological terms.

Instead, Tansley proposed that it would be more accurate to look at the natural world as a system. As opposed to the holistic values impressed upon the natural world, Tansley wrote, "the most fundamental conception is, as it seems to me, the whole *system* (in the sense of physics), including not only the organism-complex, but also the whole complex of physical factors forming what we call the environment of the biome—the habitat factors in the widest sense. Though the organism may claim our primary interest, when we are trying to think fundamentally we cannot separate them from their special environment, with which they form one physical system. . . . These [are] *ecosystems*" (Tansley 1935: 299). Tansley's idea was revolutionary because it combined in a single concept inorganic and organic components. To some, this combination may seem like the ultimate holistic conception, but it is the systems mentality that separated Tansley from his organicist colleagues. Instead of viewing the ecosystem as a single entity, Tansley followed the lead of engineering and physics, viewing both organic and inorganic parts as equal, working components of the ecosystem. Such a mechanistic perspective was intended by Tansley to open new doors for ecologists, shifting the longtime emphasis on organisms to a more equal weighting of the biotic and abiotic parts of the environment. Implicitly, in his critique of Phillips, Tansley was asserting that we need to learn more about the workings of nature before we can propose such grandiose, philosophically pleasing holistic schemes. The ecosystem concept—although seemingly holistic—was designed to help investigators reduce the workings of nature to systems that could be studied scientifically.

Some historians believe that Tansley's article marked a sharp break with the Clementsian paradigm of the community and the belief that communities behaved like organisms. As Donald Worster writes, "In [his] 1935 essay, Tansley attempted to rid ecology of all the lingering traces of organismic philosophy, expressed most recently in Clements' description of vegetation as a single living organism. . . . The often-repeated notion that the plant assemblage is more than the sum of its parts, that it forms a whole which resists reductive analysis, he took to be a fiction worked up by an overexcited imagination" (Worster 1994: 301). But in fact Tansley's article implies that he was not entirely adverse to organismic analogies for the natural world. Rather, he was angry with Phillips for misstating his position and would not support concepts that carried the organismic perspective to untenable positions. Even in "The Use and Abuse of Vegetational Concepts and Terms," however, Tansley wrote that the term *quasi organism* might be usefully applied to some mature plant communities. Taking the term beyond its logical extreme simply confused a potentially helpful analogy for aiding scientists in their understanding of how plant communities operated. While it is obvious that Tansley favored a more mechanistic approach in studying ecosystems, he still allowed for a great deal of flexibility in the definition of his new concept. For example, he noted that the boundaries of any ecosystem could be set by the scientist, depending upon the component that he wished to study. This would allow for a range of sizes of potential *isolates,* which may or may not overlap with other isolated systems. As Tansley admitted, "The isolation is partly artificial, but it is the only possible way in which we can proceed" (Tansley 1935: 300). Such freedom would still allow ecologists to look holistically at large ecosystems, and even to test whether such systems behaved like organisms. The main point of Tansley's well-argued paper was that he could not allow philosophical dogma to cloud scientific understanding, as he saw Phillips had done in his articles. Instead, he provided a concept that could be used by scientists to pry open the workings and interrelationships of the natural world. The strength of the ecosystem concept was that it could be viewed both holistically and reductively, and in the years to come, the durability that resulted from this flexibility became increasingly evident.

Applying Tansley's Concept

For the most part, ecology over the five years following the publication of Tansley's article continued in its bias toward natural history, with ecologists more apt to focus on the biotic components of a community than on the inorganic foundations and cycles of a natural system. An example of this tendency was a 1939 book by Clements and the influential animal ecologist Victor Shelford, called *Bio-ecology.* Clements and Shelford had teamed up to illustrate the importance of looking at both plant and animal communities as one biotic community. While the authors did emphasize the importance of an "exhaustive analysis of the community and its habitat" (Clements and Shelford 1939: 2), the new discipline of bio-ecology remained focused on the living constituents. The book, however, did not create the synthetic community ecology that Clements and Shelford had envisioned; as one historian observes, "Ecologists of a younger generation found the book uninspiring" (Hagen 1992: 99). Mainstream ecology was heading in a different direction.

The first ecologist to employ Tansley's ecosystem concept successfully was Raymond Lindeman, in his 1942 article "The Trophic-Dynamic Aspect of Ecology," published in the magazine *Ecology* (Golley 1993). In his analysis of a bog in Minnesota Lindeman raised basic questions on the energetics of an ecosystem, questions that would characterize ecosystem studies in future decades. Like Tansley, Lindeman began with a discussion of certain terms and concepts. Because Lindeman's work focused on the changing processes of the bog as food, and energy was transferred between different levels of the "food cycle," it is not surprising that he found Tansley's systems mentality to be most helpful. "The concept of the ecosystem is believed by the writer to be of fundamental importance in interpreting the data of dynamic ecology" (Lindeman 1942: 400).

By focusing on the energy pathways within an ecosystem, Lindeman opened a whole new landscape of ecological study opportunities. Such topics included, as ecologist Frank Golley enumerates, "the length of food chains, the efficiency of trophic transfers, the storage of energy at different levels, the rates of primary productivity, the problems of correcting energy values for losses due to respiration, predation, and decomposition, and the role of bacteria and microorganisms in cycling dead organic material" (Golley 1993: 60). The ability of Lindeman to carry out his studies was aided by the fact that his ecosystem—a bog— had clearly defined borders and a well-known successional progression. This characteristic allowed Lindeman to isolate his area of study easily—a prerequisite for the ecosystem concept that Tansley had foreseen. It was not by coincidence that Lindeman was working with the foremost limnologist in the country, G. Evelyn Hutchinson, whose published critique of Clements and Shelford's *Bio-ecology* in the journal *Ecology* had heralded a change in the direction of ecological studies. Under Hutchinson's guidance and encouragement, Lindeman defied more conservative factions and published his work. Unfortunately, he died shortly before his article went to press and so was never aware of the fact that his studies would "form . . . the cornerstone for much of post World War II ecosystem ecology" (Hagen 1992: 88).

The flexibility of the ecosystem concept created certain problems. Just as ecologists in earlier years had argued over the geographic limits of "communities," so too did critics question the boundaries of ecosystems, particularly terrestrial ecosystems. Lakes and bogs were easier to define, and for this reason early ecosystem studies were limited to these aquatic systems. Additionally, ecologists (and scientists in general) seemed confused about the difference between a *community* and an *ecosystem,* or if in fact there was any difference. Tansley had identified a community as the living components of a defined area and had intended the concept of ecosystem to include both living and nonliving constituents. But the study of communities had always implied some attention to habitat, and because Tansley had not restricted the ecosystem concept to studies of (for example) energetics or biogeochemical cycles, some ecologists began to use it simply as a new way to formulate the old concept of community (Golley 1993: 34). This development is somewhat ironic, in light of the fact that Tansley had intended to clarify concepts and terms rather than confuse them.

Nonetheless, even the study of communities was moving more toward the mechanistic, energetics-focused model that Lindeman had helped to build. Two popular texts in ecology, *Principles of Animal Ecology* (1949) and *Natural Communities* (1952), illustrate this transition. The first, a huge tome written by W. C. Allee and four highly respected animal ecologists, contained a wealth of traditional natural historical and biogeographical information on species, but also had significant sections on the influence of abiotic elements on animal populations. The second book, a text by Lee R. Dice, actually employed *community* in the sense that Tansley had intended for *ecosystem.* Instead of discussing specific biotic communities, Dice chose the adjective *natural* in an apparent effort to merge important nonliving elements into his discussion of plant and animal populations. Thus, although not all had adopted the term ecosystem, the basic concept that Tansley had posited for ecology was progressively gaining favor.

Eugene Odum and Popularizing the Ecosystem

Ecology and the ecosystem concept gained in popularity in the 1950s, when the leading voice for the young science and Tansley's concept was ecologist Eugene Odum. In the first paragraph following the introduction of his 1953 textbook *Fundamentals of Ecology,* Odum tersely laid out the ecosystem concept: "Living organisms and their nonliving (abiotic) environment are inseparably interrelated and interact upon each other. Any entity or natural unit that includes living and nonliving parts interacting to produce a stable system in which the exchange of materials between the living and nonliving parts follows circular paths is an ecological system or ecosystem. The ecosystem is the largest functional unit in ecology, since it includes both organisms (biotic communities) and abiotic environment, each influencing the properties of the other and both necessary for maintenance of life as we have it on the earth" (Odum 1953: 9). Odum went on to cover the main elements of ecosystem research—biogeochemical cycles, principles of limiting factors, fundamental concepts related to energy—as well as concepts based in population and community ecology. But by making the ecosystem the anchoring theme of his influential text, Odum ensured that the concept would be central in the training of young ecologists. Indeed, *Fundamentals* had little competition in the textbook field for many years; through its revisions, it was considered the leading ecology text well into the 1960s (Golley 1993: 67).

Odum's detailed presentation of the ecosystem concept, while still mechanistic in its perspective, was markedly holistic in its approach. He wanted ecology to establish theories that could apply generally to different ecosystems, and while reductive studies were essential in this process the ultimate synthesis had to view ecosystems as whole entities in and of themselves (Worster 1994: 364). Interestingly, Odum was trained as a zoologist at the University of Illinois (Victor Shelford was chair of the department), where he studied the heart rates of birds. He applied his knowledge about the physiology of organisms to the study of ecosystems. As Odum recalls, "The transition from bird physiology to ecosystem function was quite natural for me since it involved moving up the hierarchy from physiological ecology of populations to the physiological ecology of ecosystems. It's really not such a big step to go from whole organism metabolism to community metabolism" (quoted in Hagen 1992: 123). While Odum maintained a systems approach to studying the energetics of an ecosystem (in the spirit of Lindeman's work), it was evident that he also found many similarities between the functioning of an ecosystem and the functioning of an organism. In this way, Odum was able to find common ground for the old Clementsian organismic viewpoint and the newer, physics-inspired mechanistic perspective.

Perhaps most important, however, was Odum's foresight in emphasizing the ecosystem as the fundamental unit of ecology, since it enabled conservationists "to respond to the larger-scale problems that now require attention" (Odum, quoted in Worster 1994: 364). The deteriorating condition of the natural world was a growing concern for many in the 1950s, and by the 1960s environmental issues had become a popular cause. Major funding for ecological research in the 1950s came from an unexpected source: the Atomic Energy Commission (AEC). Odum himself received large grants to study the impact of radiation on natural systems. By 1958, the AEC-supported study in radiation ecology based at Oak Ridge National Laboratory in Tennessee was being touted as the "largest single ecological research enterprise in the United States" (McIntosh 1977: 361). Not only did this financial support contribute to the increasing popularity of ecology, it also marked a growing awareness of the interdependence within ecosystems, of which humans were a part. Whether the focus was radioactive materials, pollutants from industrial sources, or pesticides from agricultural fields, Americans were becoming conscious of the fact that their activities were impacting the natural world enough to cause repercussions for human society.

Under these conditions, the ecosystem concept "became established as a scientific paradigm in ecology" (Golley 1993: 104). The mechanistic approach was perfect for tracking how pollutants moved through natural systems, but Odum's emphasis on holism also allowed people to view ecosystems as an entity whose functioning could be damaged. As a result, conservationists were able to seize upon the ecosystem as a fundamental unit for nature preservation. Given the strength of the scientific standing of the concept of ecosystems and the obvious importance of healthy ecosystems to human well-being, people concerned with protecting nature could use the concept as a valuable tool. Ecosystems were no longer simply objects of scientific investigation; they also represented threatened entities that demanded protection.

Predecessors to Protecting Ecosystems:
Early Federal Land Protection Efforts

Although the term ecosystem was not introduced until 1935 and the protection of ecosystems (by name) did not become a popular issue until the 1960s, there were numerous predecessor preservation efforts involving large areas of land in the United States. Most notable were the federal activities in the late nineteenth and early twentieth centuries. In these efforts, little (if any) attempt is made to identify target conservation units for their ecological values. Instead, aesthetic, recreational, utilitarian and cultural values were most often the motivations. While such set-asides were specifically intended to protect scenic wonders, natural "playgrounds," and forest, water, and wildlife resources, the millions of acres reserved by the government certainly served to protect numerous ecosystems in their entirety, and the federal system that began well over a hundred years ago still provides an important foundation for protecting the variety of ecosystems in the United States. While the federal government has numerous designations for protected land, three of them—national parks, forests, and wildlife refuges—are particularly pertinent because of their importance in protecting the values inherent in the biodiversity concept.

The first significant federal land protection efforts did not involve the protection of extractable resources, as one might expect. Instead, the aesthetic, cultural, and recreational appeal of the rugged western landscape first inspired Congress to set aside parkland. While Yellowstone Park holds the distinction of being the first national park (1872), the roots of the ideals behind the Yellowstone designation lie in the 1864 law that granted the Yosemite Valley to the state of California under the condition that the state retain the area as a park for "public use, resort and recreation . . . inalienable for all time" (in Runte 1987: 29–30). Undoubtedly, in setting aside Yosemite Congress was responding to what historian Alfred Runte calls the "monumentalism" of the Yosemite landscape. The park's boundaries were set to encompass only the grand cliffs and surrounding mountains that towered above the valley, as well as a southern reserve for the largest of the huge sequoia trees in the Mariposa Grove. There was no concern for protecting the various animal and plant populations, nor for respecting watershed boundaries. The purpose of the reserve was "strictly scenic" (Runte 1987: 29).

But in this scenic value it is important to recognize the identification of the landscape's cultural significance. Well before 1864, the nation had looked to claim a heritage in its dramatic and sublime landscapes. From Thoreau's fascination with Mount Katahdin, to numerous people's rhapsodic descriptions of Niagara Falls, to the Hudson River school's distinctively American paintings, Americans were searching for a history that would give their young country an identifiable foundation. Lacking the centuries of art and architecture that their cultural relatives in Europe boasted, the people of the United States turned to their most abundant resource—land—and began to revere the most dramatic examples of natural wonder. In this way, the movement to create national parks was as much about cultural value as it was about aesthetic and recreational value.

Because Yosemite was created as a state park, and because its land area is relatively small, the founding of Yellowstone in 1872 is generally considered the birth of the national parks vision, not only in this country but in the world. The story of how the idea for a national park arose is a familiar part of conservation folklore. As many histories recount, in 1870 the members of the Washburn-Doane expedition, after completing an exploration of Yellowstone to confirm tales of the region's scenic curiosities, were sitting around a campfire discussing the various purposes the area might serve and how the men of the expedition might personally profit. But after considering different exploitative schemes, as historian Richard West Sellars writes, "in a moment of high altruism, the explorers agreed that Yellowstone's awe-inspiring geysers, waterfalls, and canyons should be preserved as a public park" (Sellars 1997: 8). A year and a half later, Congress passed the 1872 legislation, declaring that Yellowstone was "dedicated and set apart as a public park or pleasuring ground for the benefit of the people." The secretary of the interior was directed to propose regulations "for the preservation from injury or spoliation, of all timber, mineral deposits, natural curiosities, or wonders within said park." In addition, these resources and scenic wonders were to be maintained "in their natural condition" (in Runte 1987: 46–47).

While we may celebrate this act of "high altruism," one part of the

Yellowstone story is often omitted. Certainly, as with Yosemite, Congress was affirming the aesthetic, recreational, and cultural values of the American landscape. But it is important to note that the Washburn-Doane expedition was funded in part by the railroad interests. Jay Cooke, who was the chief financier for the Northern Pacific Railroad Company, saw an opportunity in the scenic value of Yellowstone. He had met with politician Nathaniel Langford—one of the more influential members of the Washburn-Doane expedition—just a few months before the exploration of Yellowstone. After the expedition, with Cooke's financial support, Langford toured East Coast cities, lecturing on the unparalleled and unique beauty of Yellowstone and promoting the interest of potential travelers. Northern Pacific also paid artist Thomas Moran to accompany a geological survey expedition in 1871. With the railroad as his patron, Moran sketched the sites of Yellowstone, and his artwork was sent to the Capitol building when Congress debated the Yellowstone bill in 1872 (Sellars 1997: 9).

In short, the railroad knew that it would only be a matter of time before their service stretched across the Montana territory. Cooke recognized that people would need a tempting incentive to leave the comfort of the East Coast to visit the Wild West. With the confirmation of the geological marvels in Yellowstone, Cooke and Northern Pacific saw that tourism was the best way to increase the traffic on their railroad. They lobbied Congress resolutely, and the government, already interested in moving people west, was easily convinced. Thus, the economic value of the scenery of protected areas was a significant influence in the genesis of the national park idea. As Sellars observes, "Indeed, in historical perspective, the 1872 Yellowstone legislation stands as a resounding declaration that tourism was to be important in the economy of the American West" (Sellars 1997: 10).

By 1916, when Congress officially created the National Park Service, a total of fifteen national parks had already been designated. The railroads had vigorously lobbied for some of these, like Sequoia, Mount Rainier, and Glacier. To lure tourists, some development in the parks and on surrounding lands—roads, hotels, and concessions—was encouraged. The railroads often took advantage of these opportunities by funding developments in order to ensure that a variety of attractions were available to their customers.

But other interests besides those of the railroad had strong influence in the designation and management of the parks, and the railroads did not always get what they wanted. In the 1880s, the railroads and mining companies proposed that an extension of the railroad be built inside the northern part of Yellowstone Park. This idea was quashed by Congress, an event that Sellars interprets as recognition of "the importance of both the park's wildlife and its wild lands—thus moving beyond the original, limited concern for specific scenic wonders of Yellowstone" (Sellars 1997: 15). This "emerging interest in protecting wilderness" was also evident in the creation of a park around Yosemite Valley in 1890. With the most scenic features already protected in the park, there was little reason to protect the more remote and seemingly less valuable lands outside of the valley. But the preservation of the High Sierra country was secured by the hard work and eloquent voice of John Muir, who had an unusual ability of convincing Washington decision makers that wilderness had value. Once again, Sellars sees this as an important development in the parks idea, calling it the "most prominent juncture between the parks movement and intellectual concerns for nature's intrinsic values and meanings" (Sellars 1997: 15). While protection for ecological purposes was evidently not an issue in this early interest in wilderness, the urge to preserve a region in its natural condition for reasons other than appreciation of its scenic wonders represents an expansion in the values that were recognized as important in setting aside land.

In 1916, the idea of preserving natural conditions was codified in the Organic Act for the National Park Service, which contained this statement of purpose: "To conserve the scenery and the natural historic objects and the wildlife therein and to provide for the enjoyment of the same in such manner and by such means as will leave them unimpaired for the enjoyment of future generations" (TNC 1975a: 264). While "unimpaired" could mean the same thing as preserved natural conditions, the language in fact does not exclude the development of certain park resources, particularly those that might help to further "the enjoyment of future generations." The phrase has become a major point of contention for those who today believe the National Park Service is failing in its mission to preserve nature. But it is evident from sections 3 and 4 of the Organic Act (which spelled out certain allowable development activities) that Congress never intended the parks to be wilderness preserves. Instead, the values inherent in the idea of a "public pleasuring ground"—where people could come and marvel at America's wonders (preferably traveling by train)—are what most influenced the early years of the national park movement.

The U.S. Forest Service, in contrast to the Park Service, had a much different objective in setting aside land. From its origin, the Forest Service's goal was to protect and manage the land and its resources for their utilitarian value. There was an Office of Forestry in the Department of Agriculture as early as 1875, but it was not until 1891 that Congress, through the Forest Reserve Act, gave the president the authority to designate land as "forest reserves." President Benjamin Harrison responded to the new law by setting aside nearly thirteen million acres by 1893. There were, however, no clear plans for the reserves and how they would be managed. After Grover Cleveland had added over twenty-five million acres during his second term, Congress passed the Organic Administration Act of 1897, establishing the purpose of the reserves as to secure "favorable conditions of waterflow, and to furnish a continuous supply of timber for the use and necessities of citizens of the United States" (TNC 1975a: 138). It is interesting to note that the forests were considered as much a source of water as a source of wood in these early years. For this reason, the boundaries of the lands set aside as reserves often extended to encompass whole watersheds. While this practice was obviously based once again in the utilitarian values of the forest, it also served to protect entire ecosystems much more effectively than the smaller and more scenically inspired national parks. While ecosystems were not a conscious part of the planning process for creating a system of national forests, the early conservation leaders did look to provide for a variety of uses of the public lands and therefore sought extended boundaries for reserves. One of the most prominent conservationists of the time was Gifford Pinchot, a German-trained forester and close friend of Theodore Roosevelt, whose often-repeated mantra, "the greatest good for the greatest number," became the slogan of the conservation movement. As Samuel Hays observed, conservation at this time was characterized as the efficient exploitation of the numerous resources of the land (Hays 1979). Thus, by means of his positions of power in the Roosevelt administration, Pinchot pushed to open the reserves for all kinds of uses: grazing, mining, hydroelectric power, and of course timber. This utilitarian philosophy has been a steady feature of the Forest Service since its inception, and although many environmentalists today feel that the Forest Service should focus more on the business of preservation, the doctrine of "multiple-use management" has a long and venerable history in the agency. While the heavy utilitarian flavor of the Forest Service's objectives has tempered over time, the management of extractive resources will always be part of its mission. Still, it is important to note that by lobbying for large set-asides decades ago the Forest Service protected significant areas of land from private development. Later efforts to protect different types of ecosystems relied heavily on the extensive holdings of the Forest Service. In addition, environmentalists would start to pressure the Forest Service to adapt the agency's mission from one focused on timber to one that protected the land to preserve an expanded range of values.

A third federal government structure significant to the concern for ecosystems is the National Wildlife Refuge System. While refuges ostensibly were created for the sole purpose of preserving species and their attendant values, the refuge set-asides are also significant in that they recognize the essential link between healthy wildlife populations and intact habitats. In protecting habitats, Americans were taking the first steps toward viewing parts of the natural world as an interrelated system, in which different integrated components must be maintained for the whole to function well. Arguably, the first refuges were designed to protect economic values—for example, the Pribolof Islands for fur seals in 1869 and the Afognak Island Forest and Fish Culture Reserve for salmon and other wildlife in 1891. But many of the refuges designated under Roosevelt were to spare certain species of bird from the voracious plume trade, thus affirming the importance of the humanistic, aesthetic, and ethical values of species over the economic value of the birds' feathers. The first protected area to be formally called a "refuge" was established in 1912—the National Elk Refuge in Wyoming—and while a National Wildlife Refuge Administration Act was not passed by Congress until 1966, the refuge system grew steadily over the years. By 2006, wildlife refuges covered more than ninety-six million acres of land.

The parks, forests, and wildlife refuges each represent the protection of certain environmental values in large land set-asides. While these efforts do not represent a concern for ecosystems per se, the concern for different attributes of the landscape—whether it was the aes-

thetic quality of the scenery, the economic usefulness of particular natural elements, or the place's value as a haven for wildlife endangered by
exploitive behavior—inspired people to reserve significant acreages
and thus protect the numerous values now associated with a variety of
ecosystems. But these early federal efforts, while diverse in their goals,
never expressly voiced a concern for protecting *diverse* land types.
They achieved protection for many different kinds of ecosystems simply by the nature of their differing objectives and by the grand style
with which huge areas were reserved. In fact, it was not the federal government that initiated specific efforts to conserve the full variety of
land types. Rather, not surprisingly, the scientific community first expressed the wish to maintain representative examples of "natural areas" that could be used as outdoor laboratories. This expression of the
scientific value of natural areas is perhaps the most significant precursor to the later concerns for ecosystem diversity.

Protecting Natural Areas for Ecological Research

In 1915, ecology was still a relatively unknown scientific discipline, but
there were enough ecologists around the country to form the Ecological Society of America. The first president and chief instigator of the
new organization was Victor Shelford, an animal ecologist from the
University of Chicago. With fifty scattered but enthusiastic members,
the group's stated purpose was "to promote the scientific study of organisms in relation to the environment and to facilitate an exchange of
ideas among ecologists" (Behlen 1981: 7). But Shelford had a more
specific agenda in mind. In 1917, he proposed that the society form the
Committee on the Preservation of Natural Conditions; the members of
the society agreed and Shelford was appointed chair. Over the next five
years, the committee worked to identify nearly six hundred areas in
North America that were "preserved or worth preserving." The slogan
on the committee's letterhead encapsulated their objective: "An undisturbed area in every national park and public forest" (Croker 1991:
124).

It is evident that it was strictly the scientific and educational value of
natural areas—as laboratories in which ecologists could study and
teach—that initially motivated Shelford and his colleagues in their
protection efforts. There were no romantic notions of protecting na-

ture for nature's sake. As Shelford had declared in an earlier text on animal ecology, "There is much sentimental nonsense about nature" (Shelford 1913: 8). The preservation of natural conditions was simply an essential component of conducting accurate ecological studies. A book that Shelford edited, *Naturalist's Guide to the Americas* (1926), was an annotated anthology documenting the areas that his committee believed deserved protection. As Shelford wrote, "A branch of biological science which obtains its inspiration in the natural order in original habitats must depend upon the preservation of natural areas for the solution of many problems" (Shelford 1926: 3). Shelford and his colleagues recognized that natural areas were being modified by humans at an increasingly alarming rate. If there was to be anything left of the original biota to study, the scientists had to take matters into their own hands.

Shelford was not alone in his concern, and other scientists added their support in speeches and articles. In a 1921 article published in *Science,* entitled "The Responsibility of the Biologist in the Matter of Preserving Natural Conditions," F. B. Sumner wrote, "It is my hope that more of our leaders in science will be aroused to the necessity of becoming also leaders in the conservation movement" (Sumner 1921: 39). Not only did Sumner offer a call to action, but he also proclaimed that "most of all, perhaps, the influence of the biologist is needed in counteracting the dominant utilitarian or materialistic trend of the day" (Sumner 1921: 41). Here was a recognition of the clash of values that scientists who studied the natural world were facing; in order to protect what they believed was important about undisturbed areas, the scientists would have to compete with powerful alternative concepts.

Other articles published in the early 1920s reveal that there were also many natural resource managers—foresters in particular—who were suggesting the need for protected areas as "yardsticks" for comparing the effects of certain management regimes. With forestry still a young and growing profession in the United States, foresters were learning about how North American natural communities would respond to silvicultural applications. For this reason, many foresters believed it was essential to maintain some forests in their original condition as a scientific control for later research. In a 1922 article from the *Journal of Forestry,* "Reserved Areas of Principal Forest Types as a Guide in Developing an American Silviculture," W. W. Ashe suggested that representative samples of different forests should be preserved: "It is believed (1) that the original forest types should be carefully mapped; and (2) that characteristic areas, *vestigial units,* of each type should be held for reference and as guides for future work" (Ashe 1922: 277). In addition, Ashe pointed out that the national parks, while more preservation-oriented than the national forests, do not cover a wide enough variety of natural areas to serve this purpose. Using the example of forest types in the Appalachians, he wrote, "If such areas are to be preserved it largely must be through the medium of National Forests" (Ashe 1922: 283). Others, such as G. A. Pearson in his 1922 article "Preservation of Natural Areas in the National Forests," echoed Ashe's idea. As Pearson observed, nothing short of "formal withdrawal of specific areas under specific provisions" seems appropriate in efforts to protect such "vestigial units" as Ashe mentioned (Pearson 1922: 285).

Wilderness

In addition to the scientific and professional concerns for protecting natural areas as vessels of valuable information, in the 1920s a new movement was gaining momentum that saw undisturbed nature as having significant recreational value as "wilderness." Such value was different from the scenic value protected by the national parks; proponents believed that the human experience of "primitive" conditions was not only healthful but also culturally and perhaps spiritually instructive. One of the earliest advocates of this view was Aldo Leopold, who, stationed as a young forester in the Southwest, published an article in the *Journal of Forestry,* entitled "The Wilderness and Its Place in Forest Recreational Policy" (1921). Noting that one of the initial arguments against the formation of national forests was that they would simply sit unused and "remain wilderness," Leopold wrote that Pinchot had promised that the forests would be developed under the doctrine of "highest use," applying the criterion of "the greatest good for the greatest number" (Leopold 1921: 718). But Leopold identified "a recent trend in recreational use policies and in the tone of the sporting and outdoor magazines" that indicated to him that those managing national forests must, under Pinchot's doctrine, provide areas for a certain kind of recreation: "Sporting magazines are groping toward some logical reconciliation between getting back to nature and preserving a little nature to get back to. Lamentations over this or that favorite vacation ground being 'spoiled by tourists' are becoming more and more frequent. Very evidently we have here the old conflict between preservation and use, long since an issue with respect to timber, water power, and other purely economic resources, but just now coming to be an issue with respect to recreation. It is the fundamental function of foresters to reconcile these conflicts, and to give constructive direction to these issues as they arise" (Leopold 1921: 718–719).

Leopold's solution was to set aside wilderness conservation areas: "A continuous stretch of country preserved in its natural state, open to lawful hunting and fishing, big enough to absorb a two weeks pack trip, and kept devoid of roads, artificial trails, cottages, or other works of man" (Leopold 1921: 719). He made clear that he was proposing that only a small fraction of land be reserved as wilderness (perhaps "one in every state") and that it be land with little economic potential. Finally, he declared that "each area should be representative of some type of country of distinctive recreational value, or afford some distinctive type of outdoor life, opportunity for which might disappear on other forest lands open to industrial development" (Leopold 1921: 719). Thus, while the preservation for recreational value was different than the preservation for scientific value, the solution—to protect representative samples of America's natural heritage—was the same. The scientific and professional communities wanted a variety of protected areas for data collection and comparison, and wilderness advocates wanted land for the growing numbers of recreational enthusiasts. The practical goals of each constituent party were virtually identical, and each reinforced the other in the development of land management policies through the 1920s and 1930s.

Certainly, those pressing for protection of natural areas were a minority voice in a culture where economic value still held primacy. Arguments like those of Ashe and others—stating that natural areas would help us to learn how to exploit forest resources more efficiently—were received most favorably. Still, Shelford's Ecological Society of America committee and wilderness proponents like Leopold did make some headway in the management communities. In 1924, Leopold, then an assistant district forester in New Mexico, convinced the Forest Service to set aside 750,000 acres at the head of the Gila

River in the Gila National Forest. Although wilderness areas would not be given formal protection until the Wilderness Act of 1964, the Gila River was considered the first national wilderness area designation. Additionally, influenced by the lobbying of Shelford and his committee, the Forest Service designated the first official Research Natural Area in 1927, located (like the first wilderness area) in New Mexico. While the protection of these lands surely did not interfere with any resource extraction projects, it is still significant that the preserved area's wilderness or scientific value was officially recognized.

Preserving a "Full Sample" of Nature

By the 1930s, some members of the growing Ecological Society of America had become uncomfortable with the active advocacy that characterized Shelford's Committee on the Preservation of Natural Conditions. Many questioned how confrontational an objective scientific organization ought to be. In 1931, the society voted to divide the committee's duties into two bodies. Shelford's committee would retain its active role in "writing letters and resolutions to government bureaus and attending hearings," while a new committee, the Committee for the Study of Plant and Animal Communities, would be the "fact-finding body." Interestingly, the new committee's letterhead had a similar message as that of Shelford's committee, but the sentiment was stated in apparently more scientifically acceptable language: "A nature sanctuary with its original wild animals for each biotic formation." While the desire to have a variety of undisturbed samples of land was unquestionable, the message of the society to Shelford was clear: his lobbying of legislators was considered a potential detriment to the group's reputation (Croker 1991: 127).

Throughout the decade, interest in maintaining areas for scientific and management studies continued to be voiced. In the late 1930s and early 1940s, numerous articles appeared in both scientific and professional journals arguing for the necessity of protecting areas as standards to compare to manipulated landscape (Hanson 1939; Piemeisel 1940; Egler 1941; Hough 1941). As one forester wrote in 1936, "Let us not be so short-sighted as to fail to profit from past mistakes, in not preserving a study area in undisturbed state in each great biotic region of the country. It is not yet too late to save a few, and far from being a luxury, it may be one of the wisest investments that the country has ever made—natural yardsticks to measure man's land management by" (quoted in Hanson 1939: 132).

Increasingly, the theme of representative samples of major "biomes" became a common concern among those interested in land protection. One author wrote of the need for "adequate examples of all major types of vegetation" (Baldwin 1941: 82). An important part of this task was to establish a classification system of vegetation types and an inventory of which types already enjoyed adequate protection. Several different classification systems were in use, many of which used the same terminology. Foresters traditionally classified forests according to "cover type." Shelford, in his 1926 *Naturalist's Guide to the Americas,* had used such broad categories in a review of each state's natural areas potential. Another example was developed by Clements and Shelford in *Bio-ecology* (1939), where they wrote of "biotic provinces." Lee R. Dice followed their lead in 1943 with his book *Biotic Provinces of North America,* in which he subdivided each biotic province into "biotic districts," "life belts," and "ecological associations." This desire to identify units of classification for natural areas was an important component of the drive to protect them, for without such a scheme there was no way of analyzing which biotic types were well represented in protected areas and which were not. Obviously, *ecosystem* had not yet become a term in general use, but the concern for different types of natural areas was certainly a predecessor to the concern for the diversity of ecosystems that would arise in later years.

A growing recognition of the importance of protecting a comprehensive sample of vegetation types is evident in a series of articles that was published in *Science* in 1941. In response to Henry Baldwin's article "An Inventory of Natural Vegetation Types and the Need for Their Preservation," Willard Van Name wrote a piece entitled "Need for Preservation of Natural Areas Exemplifying Vegetation Types." In it, Van Name heavily criticized what he saw as the paltry federal contribution to protecting natural areas, especially singling out the efforts of the Forest Service and the lack of diversity in their designations. He wrote, "We have secured really enormous reservations of the least important kinds—high mountain areas—because they are of no commercial value and nobody objects much, and we have as a rule failed entirely to protect examples of the most important of all kinds of areas and those which are disappearing most rapidly. These are the last and rapidly vanishing remnants of the various types of primeval forest" (Van Name 1941: 423).

Van Name may have been frustrated by the favorable press that the Forest Service had received for their primitive and wilderness area set-asides. Many of these, as Leopold had originally proposed in his 1921 article, were on land where profitable resource extraction was not feasible. More valuable timber lands had been left open to cutting. Van Name declared: "The most important natural areas and the most immediately threatened ones are well known now. What we need is action before it is too late" (Van Name 1941: 423). A Forest Service employee, I. T. Haig, wrote a letter for the same issue of *Science,* claiming that twenty of the major forest types were represented in forty-one areas of national forest and that the Forest Service was adequately responding to the needs of the public, the scientific community, and the natural resource professionals. Haig also chided Van Name for not being supportive of preservation work in progress, implying that the Forest Service still needed time to decide which lands were appropriate to protect (Haig 1941: 163). But Van Name's criticism had identified a weakness in the comprehensive quality of the federal protection efforts, and the urgency that he conveyed was becoming more commonplace among proponents of natural areas protection.

The Roots of the Nature Conservancy

Any talk of protecting land was virtually silenced during World War II, as the country focused on efficient output of resources for the war. In addition, Shelford's Committee on the Preservation of Natural Conditions had gradually become further marginalized after the division of responsibilities in 1931, and the Ecological Society had grown more and more wary of Shelford's activism. Even a hint of advocacy was considered a threat to the society's reputation as an objective, representative group. Finally, in July 1945, the society voted to revoke the power of the committee "to take direct action designed to influence legislation on its behalf" (Croker 1991: 143). This move confirmed fears that Shelford had voiced earlier, and the emphasis on simply "providing information" to those making decisions about land protection seemed wholly inadequate to him. By stripping the committee of any authority to influence policy, the society had unceremoniously cast aside Shelford's original vision of a protected system of outdoor laboratories defended by those who studied them.

Shelford did not take his defeat lying down, but promptly organized the ecologists who had been active on the committee and formed an independent organization that would continue the advocacy work. Called the Ecologists Union, the new group's stated objective was "the preservation of natural biotic communities, and encouragement of scientific research in preserved areas" (Croker 1991: 144–145). The formation of the organization was timely. With the war over, the need for wartime resources was past but the demand for building materials to house the new families of returning soldiers was just beginning and would continue for years to come. Fortunately, the resource management community's prewar interest in natural areas protection resurfaced in postwar years. In 1947, for example, the Society of American Foresters (SAF) created the Committee on Natural Areas, and in 1949 the society published a list of sixty-eight recognized natural areas, representing sixty-six different forest types (Buckman and Quintus 1972: i). The SAF committee also helped the new ecologists' group in locating "representative examples of old-growth forest" that were thought to be valuable examples of America's ancient landscape (Behlen 1981: 8). In this way, the management and scientific communities continued their mutual support network for establishing protected natural areas.

The work of the SAF committee was important, but the future of natural areas protection in the nongovernmental sector would lie with the seemingly modest Ecologists Union. Membership by 1949 had reached three hundred, and the union had strongly inserted itself into several political debates, including ones involving grazing lands, an international park on the Canadian border in Minnesota, and the creation of a national grassland preserve (Croker 1991: 145). In the spring of 1950, a congressman, Charles Bennett, had been enough impressed by the union's work that he proposed legislation that would create a federally supported "Nature Conservancy," modeled after the recently established British Nature Conservancy. Bennett apparently envisioned that the Ecologists Union would provide the framework for the new federal organization, assisting the Park Service and the states in preserving natural areas "having scientific, educational, or aesthetic significance" (Behlen 1981: 8–9). Bennett's bill never made it to the floor for a vote.

With the conservancy bill defeated, the Ecologists Union decided to take matters into their own hands. As the directors tried to attract a broader membership, it became apparent that the organization's name had some drawbacks. Not many at this time understood what an ecologist was, and the term *union* had some controversial associations. In September 1950, the group formally changed its name to the Nature Conservancy (TNC). By dropping ecology from their title, the advocacy group further distanced itself from the more objective Ecological Society of America and simultaneously became associated with their popular British counterparts (Behlen 1981: 10). Aided by a spotlight article in the *New York Times,* TNC began to increase its membership dramatically. As the *Times* pointed out, there was a new sense of urgency in the protection of natural areas. Certainly, the original goals of protecting different habitats and plant communities for their scientific and educational values were still primary. But there was also an important emphasis on the preservation of variety: "In addition to their obvious use in the study of basic biological research, these virgin tracts of forest desert, prairie, mountain, swamp . . . will one day be of incomparable interest as the sole survivors of the America that was. The important thing is to save as many of the diverse types as possible—now" (in Behlen 1981: 10). The concern for diversity was forcefully stated, and the significance of protecting the nation's natural heritage was evident.

Within a year the membership had doubled to 521, and by 1953 had doubled again to over 1,000. With the new support and influx of resources, the leadership decided to start acquiring land using the organization's own funds. In 1955, TNC purchased its first preserve, a sixty-acre area along the Mianus River Gorge in New York State. This method of protecting land turned out to be popular with TNC's members, likely because the membership could see actual protection occurring with the money they had given.

Through the 1950s and 1960s, acquiring land became one of the primary objectives of the new environmental group, and TNC continued to purchase what they considered to be "significant" natural areas, not only for their scientific value but also for aesthetic, ecological, recreational, and cultural values. Over these years, the organization evolved into what many considered to be the most effective and broadly popular environmental group whose main purpose was to protect land. However, it is important to note that TNC's mission did not arise in a completely unsupportive environment. As already discussed, in the 1950s there was a surge of interest in ecology, an increasing concern over endangered species and the destruction of their habitat, and ever more writers discussing the importance of the aesthetic and cultural value of nature. These trends were not limited to the United States. Significant events were occurring internationally that would eventually grow into a multinational effort that identified the "ecosystem" as its basic unit of study and conservation. This international movement for land protection would have a lasting impact in many countries, including the United States. For example, the International Union for the Conservation of Nature and Natural Resources (IUCN) and the International Biological Programme (IBP) did work that was particularly important in the history of protecting ecosystems and their diversity.

International Efforts
The IUPN/IUCN

In 1948, representatives from 23 governments, 126 national groups, and 8 international organizations met at Fontainebleau in Paris, France, and established the International Union for the Protection of Nature (IUPN), the first global environmental consortium that identified the entirety of nature as its conservation concern. The only other international organization for nature protection at this time was the International Council for Bird Preservation, founded in 1922; but this group, as its name suggests, directed its energy at only one segment of wildlife. The idea for a world-wide union had been bandied about in international circles since the end of World War II. As early as 1944, Franklin Roosevelt suggested in a letter to his secretary of state that a conference of "the united and associated nations" concerning the "conservation and use of natural resources" might serve well as a "basis for permanent peace" (Holdgate 1999: 15).

After the war, conservationists in Switzerland had been trying to mobilize an interest in global cooperation through their Office for International Nature Protection. Finally, at a 1947 conference in Brun-nen, entitled the International Conference for the Protection of Na-ture, it was decided that the appropriate forum to organize at the in-ternational level was the annual meeting of the newly established United Nations Educational, Scientific, and Cultural Organization (UNESCO). With the British biologist Julian Huxley as UNESCO's first director-general, the climate seemed right to bring different coun-tries and institutions together to form a coalition in support of protect-ing nature, and Huxley proved to be a strong proponent able to forge the necessary compromises to establish the IUPN. As might be ex-pected, bringing different countries and conservation traditions to-gether was no small task. But the representatives at the conference were able to agree on a remarkably ambitious objective for the new or-ganization. The final wording of the IUPN's preamble clearly stated its mission: "The preservation of the entire world biotic community or man's natural environment, which includes the Earth's renewable nat-ural resources of which it is composed and on which rests the founda-tion of human civilization" (in Holdgate 1999: 33).

In 1949, the IUPN held its first conference at the temporary United Nations headquarters at Lake Success in the United States. Several themes were discussed, but the one that defined the conference was the idea of "international cooperation to promote ecological re-search." The topics of the sessions easily could have been taken from later conferences. They included "the impact of DDT and other pesti-cides ... the consequences of uncontrolled introductions of exotic species; the problems of vanishing large mammals in Asia and Africa; and the emergency action needed for preserving vanishing species of flora and fauna" (Holdgate 1999: 42). The Lake Success conference was undoubtedly a significant moment in the development of interna-tional conservation, and the IUPN showed it could represent the inter-ests of many different constituents. In addition, it was unusual that the UN would support an organization that sought to bring together both governmental and nongovernmental institutions. This diversity in membership, combined with its autonomy in action, gave the IUPN the opportunity to become a broad-based, powerful force in conserva-tion.

From the beginning, there had been a debate about the use of the word *protection* instead of *conservation*. British and American representatives in particular had wanted more emphasis on recognizing the importance of conserving natural resources; the phrase *protecting nature* carried too many sentimental connotations. In 1956, those supporting a more equal recognition of utilization won out, and the name was changed to the International Union of the Conservation of Nature and Natural Resources (IUCN).

The union delegated different responsibilities to various "commissions," each of which focused on a different element of the IUCN's global environmental concern. The most powerful commissions were those dedicated to species survival, national parks and protected areas, and the promotion of ecological study and knowledge. Because of the early success of these commissions, their objectives came to define the character of the new union. As former IUCN director-general Martin Holdgate notes, "Ecology, species conservation, and protected areas, together with education ... emerged as central themes of IUCN" (Holdgate 1999: 70). Perhaps because of a strong American influence in the development of the IUCN, these topics also paralleled the reigning interests in conservation in the United States at the time. It is likely that both the international movement and the American tradition were reinforcing one another. With the growing interest in ecology in the 1950s in the United States, combined with the increasing concern for endangered species and the loss of natural areas, the new international focus on the same topics at a global scale could only add momentum.

It is interesting that the IUCN never sought to be a popular environmental group. Broadly based, with an extensive network of experts at its disposal, the union kept its focus on a narrow selection of issues. Rather than seeking the spotlight in its defense of environmental issues, the IUCN would more likely publish informational reports that would aid the efforts of smaller, more activist groups. Because of its huge constituency, the IUCN was somewhat limited in the direct action it could take. It did, however, support the founding of other organizations that would eventually enjoy the spotlight of the international environmental community. One example of an organization concerned with protecting species was the World Wildlife Fund (WWF), which the IUCN initiated in 1961 (see chapter 3). Another example, an organization concerned with furthering ecological knowledge and protecting ecosystems, was the International Biological Programme (IBP). While the IBP was not directly established by the IUCN, its objectives, particularly in studying and protecting the natural world, seemed to parallel the IUCN's mission, and the members of the IUCN—at least initially—were enthusiastic about the possible contributions of the IBP.

The International Biological Programme (IBP)

The IBP was conceived as a biological counterpart to the highly successful International Geophysical Year (1958), during which earth scientists from around the globe worked in concert to collect information about the geological processes of the planet. Sir Rudolph Peters, an English biochemist and president of the International Council of Scientific Unions (ICSU), and Giuseppe Montalenti, president of the International Union of Biological Sciences (IUBS), proposed the idea in 1960 to the executive committee of the ICSU, ambitiously suggesting that the IBP extend over several years and have as its objective the study of "the biological basis of human welfare." The idea was received favorably, and at a planning meeting in 1961 a committee laid out three possible research areas: human heredity; plant genetics and breeding; and natural communities that were liable to undergo modification or destruction. These research topics paralleled the IBP's "three specific areas for action," which were "conservation, human genetics, and improvements in the use of natural resources" (Golley 1993: 110–111).

As it turned out, the IBP did not formally become active until 1964. There were many stages of planning and coordination with other governmental and nongovernmental organizations, of which the IUCN was only one. The goals of the research and action areas were divided into seven different "sections," each of which were directed either at measuring the productivity of natural environments or assessing the impact of human use of resources. In addition, each section had separate national programs, which, combined, represented the international program. The IBP had, as many critics would later observe, a complicated organizational structure fraught with bureaucratic obstacles. But early enthusiasm for the idea of "big biology" carried the IBP through the initial planning phases, and its ambitious research and conservation programs were set into motion.

One of the more significant contributions of the IBP to the history of ecology was the international popularization of the concept of the ecosystem. While the program was initially conceived as biological in basis, it became evident that ecological research would come to dominate the IBP sections, particularly because the scientists involved were studying such major areas and issues. To carry out such research required a unit that could be useful on a large scale. With Odum and American ecologists leading the field, and with Odum himself playing an important role in the United States' contribution to the IBP, the ecosystem became the obvious unit of choice. As Worthington noted in his history of the IBP, "It is [the] ecosystem approach which distinguishes much of the IBP research from what had dominated ecology before" (quoted in Golley 1993: 112).

It is also important to note the implicit connection that the IBP, like the IUCN, maintained between pure scientific research and conservation. From its earliest conception, the IBP was designed to amass global biological information and bring it to bear on the growing concern about the capability of natural systems to support human life. By making the ecosystem a central focus of its research, the IBP went a long way in tying the ecological concept to conservation goals. As Odum announced in 1964, the year the IBP projects finally began, a "new ecology" had arrived, and "the ecosystem was its fundamental object of study" (Hagen 1992: 122). One distinct element of this new ecology was the recognition that nature was not just an outdoor laboratory but also an intricate support system for human existence. The ecological studies of ecosystem processes made this fact more apparent to ecologists than ever before.

The IBP's Conservation Terrestrial and "Representative" Ecosystems

On the topic of protecting ecosystems, it was evident that the IUCN would play a central role in one IBP section in particular, Conservation Terrestrial (CT). The main objective of CT's research was the preservation of natural and seminatural areas, for which five reasons were stated:

1. The maintenance of large, heterogeneous gene pools
2. The perpetuation of samples of the full diversity of the world's plant and animal communities in outdoor laboratories for a wide variety of research

3. The protection of samples of natural and seminatural ecosystems for comparisons with managed, utilized, and artificial ecosystems
4. Outdoor museums and areas for study, especially in ecology
5. Education in the understanding and enjoyment of the natural environment and for the intellectual and aesthetic satisfaction of mankind (Worthington 1975b: 30)

The IUCN was uniquely positioned to help with these objectives because of the work of its own Commission on National Parks and Protected Areas. Although the commission was not established until 1960, the topic of national parks and protected areas had long been important at IUCN/IUPN conferences. In 1959, the United Nations asked the IUCN to prepare a list of national parks and "equivalent reserves." The new commission was instituted, and it immediately tackled the task not only of listing protected areas but also cataloguing them through various classification schemes. The first list of world parks was published in 1961 and became a regular fixture in the IUCN's program.

In 1962, the Commission on National Parks and Protected Areas hosted the First World Conference on Parks. The IBP had prominent representation at the conference and their influence was evident in some of the conference recommendations. One of the most telling recommendations directed the IUCN to "work closely with the IBP to bring into existence a series of Natural Reserves providing permanent examples of the many diverse types of habitats, both natural and seminatural, so as to preserve them permanently for world science" (Coolidge 1963: vii). Two other important recommendations suggested that a working group be established to develop a list of representative habitats for each main bioclimatic region and that a refuge be reserved "for every kind of plant or animal threatened with extinction." Although the term ecosystem was not yet employed by IUCN staffers, it was evident that this was the direction the terminology would take under the guidance of the CT section of the IBP. As set up, the IBP, after receiving initial information from the IUCN, was in perfect position to generate the kind of data that the IUCN needed to move on its stated objectives. The IBP gave the IUCN an additional network of researchers from which to obtain information about existing or potential protected areas. Perhaps most important to note, however, is the emphasis that both the IBP and the IUCN placed on complete representation of ecosystems. The value of diversity was central to the conservation goals of both international organizations, and the objectives that emphasized protecting the complete variety of natural systems would serve to influence future conservation efforts.

Unfortunately, it seems that the CT section served mainly as a catalyst for later smaller ecosystem conservation projects that would take place after the IBP had finished—the actual accomplishments of the IBP in protecting ecosystems disappointed many conservationists. While the grandiose plans did mobilize scientists in many different countries to work for common understanding of the earth's biological systems, the applied element of the program ended up as a secondary concern. In 1965, a year when many IBP activities were just beginning, one telling sign of the disconnection from the conservation world was that the directors of the CT section decided to distance themselves from the IUCN, causing a fair amount of consternation at the union. The reasoning of the members of the CT was that their objective—to protect land of scientific importance—was more narrow, whereas the IUCN's conservation goals were much broader. It seems that the scientific core of the IBP had become somewhat uncomfortable with the influence of conservation concerns on their research. Similar to the Ecological Society of America in its decision to marginalize Shelford's Committee on the Preservation of Natural Conditions, the international scientific community felt it needed to keep its scientific objectives clear. The IBP still contributed valuable information to the IUCN's conservation efforts, but as Holdgate notes, "with hindsight, it remains extraordinary that IUCN and the world conservation community got so little out of the conservation section of IBP" (Holdgate 1999: 95–96).

At the national level, the character of the IBP sections often became defined by the interest of the scientists in charge. In the United States, ecologists saw the availability of IBP monies as an opportunity to fund large-scale ecological research. Led by W. Frank Blair and Frederick E. Smith, the American contingent seized upon the concept of the CT section but adapted it to their own objectives, renaming it the Conservation of Ecosystems program (CE). Planning on an ambitious scale, the ecosystem ecologists initiated six separate "biome" projects to study the processes of grasslands, deserts, deciduous forests, coniferous forests, tundra, and tropical forests (Worster 1994: 372). But in addition to the obvious goals of advancing scientific understanding, the American CE program also had a primary conservation objective: "The establishment within the United States and its possessions of a comprehensive system of protected research reserves" (Darnell 1976: 105). This mission had remarkable similarities to Shelford's original vision for the Ecological Society of America. The tradition of scientists working for preservation had survived and now was even connected to a much larger international agenda.

Although it seems that the US/IBP suffered a bit from the same bureaucratic problems that plagued its international counterpart, a great amount was accomplished. Scientists gathered reams of data about the North American biomes. Equally as significant were the trends in ecological research that were established. The US/IBP had an important impact on the training of practitioners, on advances in scientific methodology, and on data-collection techniques. But most notably, they had established the ecosystem as the international unit of ecology. In addition, not only had the ecosystem come of age as a "scientific paradigm" (Golley 1993: 104), it also had been introduced into the vocabulary of global conservation. Similar to the scientific pedigree enjoyed by those working in genetics, those interested in protecting land gained credibility from the conservation linkage to science implicit in the term *ecosystem* that lifted their cause above the level of sentimental nature preservation.

Unfortunately, many believed that overall IBP activities, which officially ceased in 1972, had fallen short in their conservation objectives. But one legacy was the many systems and networks established by national IBP programs of which conservationists might take advantage. For example, in the efforts to identify natural areas around the world that deserved protection, appropriate national leaders were given "check sheets" for IBP areas designed to evaluate an area's suitability. The information collected on these sheets could have been used by groups interested in carrying on the IBP's work. In addition, while the CT section had made advances in studying different kinds of ecosystems, it had also struggled with a global classification scheme designed to identify gaps in the preservation of "representative areas." It seemed evident to the international conservation community that while the

IBP had made definite contributions, a new program was required that would pick up where the IBP left off.

The Biosphere Conference and MAB: Ecosystems, Species, and Genes

For this purpose, UNESCO held a conference in the fall of 1968 that was jointly sponsored by the IBP and IUCN, as well as by the UN Food and Agriculture Organization and the World Health Organization. The full title of the gathering was the Intergovernmental Conference of Experts on the Scientific Basis for the Rational Use and Conservation of the Resources of the Biosphere, more commonly referred to as the Biosphere Conference. As typical of such UN conferences, representatives from many countries, foundations, and intergovernmental and nongovernmental organizations were present. In their first recommendation, the conferees praised the work of both the IBP and the IUCN, suggesting that the IBP "might usefully be followed by an international programme of expanded and strengthened research, education and implementation on the problems of man and the biosphere" (UNESCO 1970: 211). The second recommendation, entitled "Research on Ecosystems," outlined a "world programme" that would in essence continue the work of the IBP, studying not only natural systems but also those modified by humans ("semi-natural, artificial, or cultural"). It was an ambitious agenda. But because the organizers of the IBP had limited their program's existence to a specific time frame, there was a desire in the international community not only to extend and institutionalize the important research that the IBP had initiated but also to improve upon the conservation efforts.

Perhaps the most important recommendation of the conference—in terms of providing a precursor to the concept of biological diversity—was recommendation 7, "Utilization and Preservation of Genetic Resources," which suggested that "special efforts must be taken urgently to preserve the rich genetic resources that have evolved over millions of years and are now being irretrievably lost as a result of human actions" (UNESCO 1970: 216). The first step in these efforts was articulated in a particularly significant way. The conference called for the "preservation of representative and adequate samples of all significant ecosystems in order to preserve the habitats and ecosystems necessary for the survival of populations of species" (UNESCO 1970: 216). Here was another concise statement, echoing the objectives of the IBP's Conservation Terrestrial section, bringing together all three levels of diversity in one expression that supported preservation: the necessity of protecting a representative sample of ecosystems, for the purpose of maintaining animal and plant species, such that the genetic heritage of the earth remain intact. Other steps further outlined tasks for protecting germplasm resources, and the importance of establishing protected areas for "remnant populations of rare and endangered species of plants and animals" was also highlighted. Thus, this particular recommendation, far from being narrowly focused on genetic resources as its title suggested, recognized the interdependent quality of such conservation efforts that were to become the hallmark of protecting biodiversity.

The recommendations of the Biosphere Conference led to the establishment in 1970 of UNESCO's Man and the Biosphere Programme (MAB). The new international effort was modeled primarily on the IBP experience but there were several significant differences. First, the practical application of the IBP studies had been widely criticized. Research had not been well coordinated among the different countries and (with the United States as a prime example) the IBP objectives in Western countries had been manipulated to match the research goals of the dominant scientists. MAB placed a much heavier emphasis on conservation and switched the focus from the developed world to developing countries, many of which had been marginalized in the IBP. Second, while MAB continued the IBP mandate of studying the productivity of all different ecosystems from the poles to the tropics, MAB also extended the program to encompass studies of human dominated ecosystems, such as agricultural lands. Third, the shift from IBP's academic roots in the ICSU to the governmental science of UNESCO assured that the applied value of any research would be carefully considered (Golley 1993: 162).

MAB's Biosphere Reserves

Like the IBP's different "sections," MAB outlined its objectives in fifteen different "projects," each representing a different research theme.

Project 8, "Conservation of Natural Areas and the Genetic Resources They Contain," came directly out of recommendation 7 at the Biosphere Conference and was closely connected to the CT section of the IBP. In 1971, the International Coordinating Council for MAB first articulated the necessity of establishing a worldwide system of reserves to protect "scientific, educational, cultural, and recreational values" (UNESCO 1973: 21). Ecosystems and their valuable components were the stated targets of the reserve systems of protected areas. The 1973 annual report of MAB's Expert Panel on Project 8 even more clearly declared, "The primary objective of such an international programme must be to ensure that adequate examples of all important and representative biome subdivisions are protected. National Committees may have to prepare detailed inventories of ecosystems of importance within the various biomes in their countries for identifying and conserving adequate representative samples of such ecosystems" (UNESCO 1973: 21). The similarity to the IBP structure was obvious. National committees would be given the task of identifying gaps in protection and would subsequently propose areas that would adequately encompass ecosystems in danger.

There was, however, one significant administrative difference in MAB's conservation efforts: the concept of the biosphere reserve. First proposed at a 1971 meeting, the biosphere reserve concept was subsequently elaborated by a task force that published its work in 1974. The task force wrote, "The objectives of the international network of biosphere reserves are: 1. To conserve for present and future human use the diversity and integrity of biotic communities of plants and animals within natural ecosystems, and to safeguard the genetic diversity of species on which their continuing evolution depends. 2. To provide areas for ecological and environmental research including, particularly, baseline studies, both within and adjacent to these reserves, such research to be consistent with objective (1) above. 3. To provide facilities for education and training" (UNESCO 1974: 6). These objectives express a concern for ecosystems, species, and genetic diversity, not only for economic and utilitarian motives but for scientific and educational values as well. Perhaps just as significant were the essential characteristics that the task force established for the selection of a biosphere reserve. Each reserve had to meet the following four basic criteria:

1. Representativeness: "A reserve should represent as many charac-teristic features of the particular biome as possible."
2. Diversity: "Representative Biosphere Reserves should contain the maximum representation of ecosystems, communities, and organ-isms characteristic of the biome."
3. Naturalness: "Samples of biomes in their natural state."
4. Effectiveness as a conservation unit: "This criterion involves a number of factors such as size, shape, and location with respect to natural protective barriers." (Risser and Cornelison 1979: 2)

With such objectives and criteria, the concept of the biosphere re-serve was one of the first to merge these previously separate concerns. Project 8 of MAB had its strongest roots in efforts to protect ecosystems and natural areas, thus taking the lead of the IBP but also building upon the tradition of IUCN's Parks and Protected Areas Commission. The interest in genetic diversity that had grown throughout the 1960s had spilled over into UNESCO's MAB planning. The concern for species, not only as valuable conservation objects in and of themselves but also as the working components of ecosystems and the carriers of genetic di-versity, had strong constituencies both nationally (with the passage of the Endangered Species Act) and internationally with the success of the World Wildlife Fund. In short, Project 8 and the biosphere reserve con-cept identified ecosystems as an essential conservation unit by recogniz-ing the broad values of the different components contained within each ecosystem. Additionally, the use of "diversity" as a key criterion served as a unifying concept that brought together the separable concerns for the natural world. This emphasis would come to characterize much of the conservation language in the 1970s.

Further International Support for Protecting Ecosystems

It is important to recognize that the biosphere reserve concept of the MAB program did not come out of a void in the international conser-vation circles. Besides the influence of the IBP, several other notewor-thy events supported the idea of protecting a range of ecosystems dur-ing the early development stages of the biosphere reserve concept. In 1972, the IUCN held its Second World Conference on National Parks, at Grand Teton National Park in Wyoming. Building upon the recommendations from the First World Conference in 1962, which had called for a system of natural reserves that would conserve "representative habitats," the first recommendation of the 1972 conference was entitled "Conservation of Representative Ecosystems," calling upon "all governments to widen the coverage of their protected areas so as to ensure that adequate and representative samples of natural biomes and ecosystems throughout the world are conserved in a coordinated system of national parks and related protected areas" (in Elliott 1972: 442). One particular ecosystem—tropical rain forests—was highlighted as a special concern in the conference's second recommendation, recognizing "the rapidly accelerating destruction of these ecosystems now proceeding in many countries and the consequent danger of extinction of plant and animal species and communities, and depletion of genetic resources" (Elliott 1972: 442). As with Project 8 of the MAB program, it was evident that conservationists at this time considered the conservation of ecosystems as bound together with efforts to conserve species and genes. The protection of natural areas had become far more detailed and ecologically informed in its objectives.

Another significant statement came from the influential United Nations Conference on the Human Environment in Stockholm in 1972. One of the hallmarks of this conference had been the emphasis placed on the importance of allowing developing countries to continue to raise their standard of living while observing environmental standards. Thus, the use and conservation of "resources" were pervasive topics throughout the list of principles and recommendations. Ecosystems were singled out as one of several key natural resources that deserved special attention. As principle 2 of the United Nations Conference on the Human Environment conference reads: "The natural resources of the earth, including the air, water, land, flora and fauna, and especially representative samples of natural ecosystems must be safeguarded for the benefit of present and future generations through careful planning or management, as appropriate" (P. Stone 1973: 148). The charge is repeated in recommendation 38, which calls upon governments "to set aside areas representing ecosystems of international significance" (P. Stone 1973: 164). While the concern for ecosystems was only one among numerous others expressed at the conference, their inclusion is further evidence of the importance placed on ecosystems conservation at the time.

One other 1972 conference, the General Assembly of UNESCO, made an important statement concerning the value of natural areas not just for their resources but also as examples of our "global heritage." By adopting the Convention for the Protection of the World Cultural and Natural Heritage, UNESCO was linking together the concepts of natural and cultural value. The convention was designed to provide "a framework for international cooperation in conserving the world's outstanding natural and cultural properties" (IUCN 1982: 3). While there were different criteria identified for selecting cultural and natural heritage sites, it was apparent that the convention was asserting that natural areas were as significant to humanity as our most cherished cultural sites. "By adopting the Convention, nations recognize that each country holds in trust for the rest of mankind those parts of the world heritage—both natural and cultural—that are found within its boundaries; that the international community has an obligation to support any nation in meeting this trust, if its own resources are insufficient; and that mankind must exercise the same sense of responsibility to the works of nature as to the works of its own hands" (IUCN 1982: 7).

It was, in essence, a kind of cultural value that natural areas acquired through association. While the World Heritage List (which was to be maintained by the IUCN's Commission on National Parks and Protected Areas) was less focused on traditional environmental concerns, the concept of world cultural and natural heritage gave the international community another motive for protecting valuable areas. It also articulated another important value—a link to human culture—that natural areas provided to society. In short, the focus on ecosystems was prevalent in international forums on conservation. Whether as areas for scientific study, for protecting endangered species and the global gene pool, or as repositories of natural or cultural legacies, the protection of a variety of ecosystems seemed to be one conservation objective with which many in the international community agreed.

The U.S. Federal Committee on Ecological Reserves

In the United States, national efforts to conserve ecosystems appeared to take the lead of the international efforts described above. Just as the IBP inspired UNESCO to establish MAB, so too did the US/IBP stimulate federal efforts to protect ecosystems. In an attempt to facilitate the conservation aspect of the US/IBP ecological research, the government instituted the Federal Committee on Research Natural Areas in 1966. The committee was made up of representatives from all of the important land-managing federal agencies and also included observers from private conservation organizations. Its three primary objectives, according to the directory of natural areas published in 1968, were:

1. To assist in the preservation of examples of all significant natural ecosystems for comparison with those influenced by man
2. To provide educational and research areas for scientists to study the ecology, successional trends, and other aspects of the natural environment
3. To serve as gene pools and preserves for rare and endangered species of plants and animals (quoted in TNC 1975a: 260)

The language and intent were very similar to that of the Biosphere Conference, which in the same year was outlining its plans for the MAB program.

In 1970, the Committee on Research Natural Areas was dismantled, largely because the Office of Science and Technology, which had served as the committee's base of operations, was abolished (Darnell 1976: 105). But in 1974, with the support of the National Science Foundation and the recently established Council on Environmental Quality, the committee was reborn, this time under the title of the Federal Committee on Ecological Reserves (FCER). The charge of the new committee was virtually identical to the old: "The establishment and maintenance of a network of protected field sites that represent a full array of the nation's terrestrial, freshwater, and marine ecosystems" (FCER 1977: 1–2). This objective included providing protection to a "range of diversity, including common, rare, and endangered species or disjunct populations" (FCER 1977: 5). The new committee emphasized the scientific importance of protecting ecosystems as "baseline areas" where scientific studies could take place and the effects of certain management regimes could be tested. But the emphasis on diversity and the inclusion of concerns for all species, both common and rare, is an indication of similarities with international conservation efforts of the time. While the administrative power of the Committee on Ecological Reserves can be debated, its 1977 *Directory of Research Natural Areas on Federal Lands* listed 389 areas covering 4.4 million acres in 46 states and 1 territory. The cooperating land agencies included the U.S. Forest Service, the Bureau of Indian Affairs, the Bureau of Land Management, the Fish and Wildlife Service, the National Park Service, the Department of Defense, and the Tennessee Valley Authority (FCER 1977: 5). It was a significant system of reserves, built with substantial interagency cooperation.

Not surprisingly, many research natural areas, besides being already protected as national parks, forests, or wildlife refuges, were also distinguished under other national systems. Some, under the Wilderness Act of 1964, were established or proposed wilderness areas. Others had been previously identified by the Society of American Foresters as SAF natural areas. Still others were already recognized by the international conservation community, either as biosphere reserves or as IBP natural areas. Thus, one criticism of this proliferation of protective designations was that the extensive overlapping was inefficient and perhaps even diminished the importance of separate conservation systems. But different designations ultimately offered different levels of protection and recognized different values represented in each unique site. Most important, the numerous methods for protecting ecosystems are evidence of the range of interests represented in the conservation world at the time. The apparent lack of coordination among those interested illustrates the dispersion of similar objectives that characterized the conservation community of the 1970s. While the FCER provided some unity for government agencies, other groups remained unorganized as a force, and the massive goal of protecting a variety of ecosystems and their components called for a unifying concept that would bring environmentalists together.

Conclusion

The concept of the ecosystem and the predecessor causes of protecting natural areas represented important attempts at identifying and preserving a collective of values. While the ecosystem was originally conceived as a framework for more reductive analysis, it grew into a holistic conception of nature that served as a bridge between the scientific and conservation worlds. The tradition of protecting land in the United States is represented at the federal level with the designations of the national parks, forests, and wildlife refuges, and while these early efforts did not specifically target ecosystems as units of conservation, they did seek to preserve the aesthetic, recreational, cultural, and utilitarian values that would come to characterize the conservation of ecosystems and biological diversity. The scientific community, led by Victor Shelford, promoted the protection of representative samples of the natural world as outdoor laboratories and classrooms. A similar impulse was represented in natural resource management by those interested in preserving "vestigial units" of nature as "yardsticks" for applied management regimes and as wilderness recreation areas.

The scientific activism evolved into the Nature Conservancy, a nongovernmental organization whose strategy of preservation through acquisition has led environmentalism's fight to protect significant pieces of land. The international movement represented in the work of the IUCN, the IBP, and MAB served to focus attention on the importance of preserving a diversity of ecosystems, and this work influenced several connected ecosystems conservation efforts in the United States. In particular, the link between protecting ecosystems and preserving species and genes was made explicit in several environmental projects around 1970.

The recognition of this connection and the importance of protecting nature's multiple hierarchical levels planted the seed for further discussions over the next decade. With the wide range of interests—both national and international—expressing concern over the variety of component parts of the natural world, it was apparent that environmentalism needed a common cause behind which to rally, one that would better coordinate and motivate the work of conservation groups. The common interest in diversity seemed to be a theme that tied environmental concerns together.

6 The General Scientific and Conservation Interest in Diversity, 1950–1980

In addition to the historical development of concerns for the different hierarchical levels of biological diversity, there was a significant discussion in the 1950s, 1960s, and 1970s regarding the role that diversity played in the function and stability of an ecosystem. This discussion began for the most part in ecological circles and centered on the controversial but widely accepted theory that diverse systems were more stable than simple systems (that is, if one reduces an area's biological diversity, one risks crippling the functioning of the ecosystem). The apparent connection between diversity and stability filtered into the popular environmental vocabulary, where it has become one of the most important tenets of biodiversity conservation. Even though scientists never could agree on whether a consistently measurable ecological relationship between diversity and stability exists, conservationists stand behind the claim that a decrease in diversity will likely result in the "unraveling" of ecosystemic processes. Paul Ehrlich's rivet-popping analogy (see chapter 1) is perhaps the best example of this belief.

But well before Ehrlich introduced his analogy, the conservation world had begun to view diversity in general as a quality that was worth preserving. The protection of variety, for many in the 1960s and 1970s, served as an overarching collective cause that could encompass many of the concerns about human impact on the environment. An excellent example of the expression of this viewpoint is a 1965 conference held in Warrenton, Virginia, entitled Future Environments of North America, at which prominent conservationists of the day convened, presented papers, and carried on a lively debate about the value of diversity. By the 1970s, diversity had begun to work its way into the language of environmental groups, especially those interested in protecting species and ecosystems. To illustrate this point, this chapter looks closely at changes during the decade in the institutional language of the Nature Conservancy, and in particular at a TNC-backed legislative bill, the Natural Diversity Act. The interest in diversity, like the environmental concerns present in conservation at the time, was broadly based but still amorphous in shape. It would take a well-articulated concept—introduced at the appropriate time and place and incorporating multiple concerns—to bring various conservation constituencies together. Ultimately, this is what the concept of biological diversity provided.

While scientific interest in diversity can be traced back a number of centuries, at least to the time of Linnaeus, the modern discussion about diversity's role in the stability and functioning of an ecosystem has its roots in works first published in the 1950s. But before examining the scientific literature, it is helpful to review the publication that many consider to have inspired interest in the scientific questions about the ecological relationship between diversity and stability and in the conservation implications of the loss of diversity.

Charles Elton and the Conservation of Variety

In 1958, British ecologist Charles Elton published the highly regarded *Ecology of Invasions by Animals and Plants,* arguably the first book to articulate the conservation implications of a decrease in the diversity of natural areas. The intention of the work, in Elton's own words, was "to bring together ideas from three different streams of thought with which I have been closely concerned during the last thirty years or so . . . faunal history, . . . ecology, . . . [and] conservation (Elton 1958: 13). As the title implies, Elton was most interested in the ecological impact of exotic species on local environments: introductions, or "invasions," that were usually facilitated, if not initiated, by an increasingly mobile human population. The body of the book discusses the flora and fauna of the different continents, the breakdown of natural barriers by human travelers, and the changes in the ecological mechanics when new species are introduced. In the final two chapters, however, Elton turns to conservation, first describing his rationale for why conservation is important, and then in the final chapter discussing what he believed should be the ultimate goal of conservation: "Keeping or creating sufficiently rich plant and animal communities in our changing landscape—that is . . . conserving ecological variety" (Elton 1958: 153). This closing chapter, "The Conservation of Variety," is often referred to as one of the first expressions of a general concern for diversity in the natural world.

Elton, as discussed earlier, was one of the more influential ecologists in the 1920s and 1930s. His 1927 work *Animal Ecology* introduced several important ideas into ecology, including the concepts of the food chain and the niche. Also, Elton's discussions of "producers" and "consumers" helped to support the more mechanistic approach to studying nature, in which different components of a natural system played specific roles in its overall economy. His ideas were particularly useful in the concept of the ecosystem introduced by Tansley in 1935. But unlike some of his more scientifically objective colleagues, Elton showed an interest in conservation from early in his career. Donald Worster identifies as one source of Elton's environmental concern his meeting Aldo Leopold in 1931 at the Matamek Conference on Biological Cycles. As Worster writes, "That meeting was apparently important for both men; soon thereafter Leopold became a convert to the ecological view of nature, and Elton then began quoting Leopold on the need for a 'conservation ethic'" (Worster 1994: 300). In the 1940s, Elton helped to form the British Nature Conservancy and was active in lobbying for land protection (Takacs 1996: 24). By the time he published *The Ecology of Invasions by Animals and Plants* (1958), he had amassed a broad experience, traveling around the world studying the earth's biota and witnessing firsthand the effects of an expanding human population on an increasingly disturbed natural world.

Elton's decision to frame his discussion in terms of species "invasions" was remarkably prescient. The spread of exotic species and their sometimes destructive impacts have since become major issues for biodiversity conservation and efforts to protect native flora and fauna. Elton was the first to discuss the topic at length in a book and connect it directly to conservation concerns. His detailed examples— including the chestnut blight, the sea lamprey, and the European starling—became frequently repeated stories of the cost of conscious and unconscious species introduction. Elton thus provided the ecological community with a disturbing new perspective of human impact on the natural world. He also dedicated a major part of the book to describing the acute effects of alien species on island biota, observations that fed into the growing interest in population biology and the related theory of island biogeography.

But it is Elton's concern for the "conservation of variety" that is most significant. Throughout the book, Elton used examples of how landscapes simplified by human civilization had invited invasions and "explosions" of certain species. Agriculture and natural resource extraction left the land simplified, causing "a decrease in richness and variety of species," making impacted ecosystems vulnerable to wild cyclical fluctuations. It was not just human-modified landscapes that were subject to these disturbances, Elton pointed out: "Some invaders are also penetrating the more stable and mature communities of ocean and natural forest," disrupting the established ecological relationships and driving native species out of ecosystems (Elton 1958: 154). It was this decrease in diversity and the subsequent loss of stability that worried Elton most. "If the wilderness is in retreat, we ought to learn how to introduce some of its stability and richness into the landscapes from which we grow our natural resources" (Elton 1958: 155). Maintaining variety was the key to maintaining stability. In this claim lay the basis of one of the most popular and enduring conservation arguments of recent decades: diversity is important because complex ecosystems are more stable than simplified ecosystems. Diversity begets stability.

Elton, being a consummate scientist, was careful to note that his theory—that "the balance of relatively simple communities of plants and animals is more easily upset than that of richer ones; that is, more subject to destructive oscillations in populations, especially of animals, and more vulnerable to invasions"—was not yet proven. "Much more extensive analysis and discussion" would be required to confirm it (Elton 1958: 145, 151). He was clear, however, that he saw the protection of variety as essential and his ideas about diversity's connection to stability as easily defensible: "It is by no means a far-fetched idea, and even if it seemed so we should still need to explore it by research, because the whole matter is supremely important to the future of every species that inhabits the world" (Elton 1958: 146).

Diversity-Stability in the Scientific Literature:
Foundational Ideas

By the time Elton had published his book, a discussion within ecology about the relationship between diversity and stability had already begun. In the 1950s, ecology as a discipline was enjoying a surge of popularity and had attained a critical mass of trained ecologists, teachers, and students, such that it was possible to attempt syntheses of the previously scattered literature, organizing theories and concepts into more generalized textbooks (Odum 1953: v). The first half of the century had seen the development of many influential, sometimes competing, ecological concepts: Cowles's work on succession, Clements's idea of climax, Gleason's individualistic theory of plant association, Tansley's concept of the ecosystem. In 1953, Eugene Odum published his landmark textbook *Fundamentals of Ecology,* in which he laid out basic principles in an organization "without precedent," an organization that he labeled as "frankly something of an experiment" (Odum 1953: vi). As discussed in chapter 5, his experiment apparently was well conceived, for his text was popular and would influence many books and articles on ecology that appeared in the later 1950s and early 1960s.

Besides Odum's role in popularizing the use of the ecosystem concept, his extended definition of ecology also revealed the roots of the diversity-stability debate that emerged later in the ecological literature. A common way to define ecology was as "the biological science of environmental interrelations," but Odum proposed something slightly more detailed: ecology was "the study of the structure and temporal processes of population, communities, and other ecological systems and of the interrelationships of the individuals composing these units" (Odum 1953: 4). By spelling out that ecology was about both *structure* and temporal processes (or *function*), Odum was making clear that ecology brought together two approaches that traditionally were separate in biological studies (Kormondy 1965: 211). This was an issue upon which Odum would expand in later editions of his text (1959, 1971) and in articles like "Relationships between Structure and Function in Ecosystems" (1962). The issues that were to emerge in the literature were indicated by this title: Just what was the relationship between structure and function in ecosystems? How did the composition of the biological community impact functions like energy flow and nutrient cycling? In trying to answer questions like these ecologists first became formally interested in the role of the entirety of diversity in ecosystem processes. It was not just the study of diverse organisms themselves; nor was it simply the study of the energy and material functioning of an ecosystem. It was the analysis of how the two worked together.

One central concept that grew from these questions about structure and function was that diversity had a direct and positive impact on the "stability" of an ecosystem. This principle was difficult to prove, partly because of the difficulty of defining both "diversity" and "stability." But this problem did not stop many eminent ecologists from creating hypotheses that ultimately pointed toward the conclusion that increasing diversity played a major role in ecosystems achieving stability. One of the first publications to address the theory directly was Robert MacArthur's "Fluctuations of Animal Population and a Measure of Community Stability," published in *Ecology* in 1955. MacArthur was interested in how the structure of food webs influenced the stability of populations. Citing his observations that "stability increases as the number of links [in a food web] increases" and that relatively simple arctic ecosystem populations fluctuate far more widely than more diverse tropical ecosystem populations, MacArthur proposed that an equation measuring trophic diversity could be used to describe community stability (MacArthur 1955: 535). This equation was based on a function published by Claude Shannon and Warren Weaver in 1949 and came to be known as the Shannon-Weaver diversity index (Goodman 1975: 239). MacArthur's conclusions were largely intuitive, and the use of the Shannon-Weaver equation did not represent any mathematical proof of a connection between diversity and stability. He did succeed, however, in putting into words a relationship that seemed logical to scientists, though it was one that had never successfully been proven.

Another seminal article was G. Evelyn Hutchinson's "Homage to Santa Rosalia; or, Why Are There So Many Kinds of Animals?" published in the *American Naturalist* in 1959. Hutchinson took MacArthur's ideas even further, proposing that because diverse communities were more stable, evolution would favor those ecosystems with high levels of diversity and hence diversity would increase over time, providing the stable conditions under which a wide range of species could develop. This idea was strongly supported by theories of ecological succession, which emphasized that ecosystems advanced toward a more complex, diverse state until they achieved a mature stability. Four years later, Ramon Margalef restated Hutchinson's evolutionary perspective, in his article "On Certain Unifying Principles in Ecology," also published in the *American Naturalist.* He proposed that scientists define the maturity of an ecosystem largely in terms of diversity and energetics: "An ecosystem that has a complex structure, rich in information, needs a lower amount of energy for maintaining such a structure" (Margalef 1963: 361). Diversity, it seems, was both indicator and agent of how stable and therefore how mature an ecosystem might be.

Margalef's interest in unifying principles, in particular the formal rules defining what constituted and permitted "balance" in nature, was apparently shared by a number of his colleagues. In 1967, MacArthur and Wilson published their landmark study *Theory of Island Biogeography.* As discussed in chapter 4, the two biologists tested the hypothesis that the diversity of species on islands was directly related to land area and the distance that a species had to travel to colonize an island. In essence, diversity was regulated by certain geographic factors that helped to maintain a balance between the stability of the island ecosystems and the variety of species. This concept has played a major role in criticisms of land fragmentation by development and the subsequent loss of species diversity.

Several other articles published in the late 1950s and early 1960s addressed different theoretical aspects of the diversity-stability relationship, but none were able to come up with satisfactory scientific evidence of its existence. However, most seemed to believe that it was only a matter of time before the theory was proven. Odum's text *Ecology* (published in 1963, a more simplified treatment than *Fundamentals*) had a brief section called "The Importance of Diversity," in which he wrote: "It is now generally assumed, but without much real scientific evidence, that the 'advantage' of a diversity of species . . . lies in increased stability" (Odum 1963: 34).

It should be noted, as is evident from the above-cited studies, that this debate focused almost exclusively on *species* diversity and initially paid little attention to *genetic* or *ecosystem* diversity. But as the discussion progressed, scientists were more apt to expand their definition of diversity, acknowledging that species diversity, although the first order of concern, obviously depended extensively on other levels of diversity. In addition, biologists and ecologists began to ask more probing questions about how to measure and define diversity and stability. Was diversity simply the numbers of different species, or did the relative abundance of different organisms matter? Was stability only population fluctuation, or were other factors important to consider? The more detailed analyses of the relationship between diversity and stability primarily confirmed that the issue was far more complex than conceived in the earlier articles.

The 1969 Brookhaven Symposium

By 1969, the discussion of diversity's functional role had attracted enough interest in scientific circles so that Brookhaven Biological Laboratories dedicated their prestigious annual symposium to the topic. The symposium was called Diversity and Stability in Ecological Systems. The chairs, G. M. Woodwell and H. H. Smith, declared in the preface to the symposium's proceedings a surprisingly practical objective, in addition to the purely scientific interest of advancing the understanding of ecology: "A major means for assuring the continuity of life appears to be the number of species per unit area, diversity. . . . While the success of evolution in accommodating to the extremes of conditions on earth and even to man is clear, the continual success of man-dominated systems which have many parallels with natural ecosystems is much less clear. The objective of this Symposium was to examine the meaning of 'diversity' and 'stability' as used by students of natural systems and to explore relationships to man-dominated ecosystems" (Woodwell and Smith 1969: v). Echoing the warnings of Charles Elton eleven years before, Woodwell and Smith expressed the scientific-environmental concern that had become popular over the previous decade. In ecosystems where humans had significant impacts, such as areas affected by agricultural and commercial development, it was apparent that the ecological processes of such sites were compromised. By focusing on diversity and stability, the 1969 Brookhaven Symposium gave credence to the idea that studying these two qualities of natural systems might help us understand the extent and severity of human transgressions.

While a number of articles in the proceedings addressed the practical issue articulated by the chairs, the majority of the papers were primarily concerned with exploring the nagging scientific questions that surrounded the diversity-stability relationship. In particular, the definition and measurement of stability was perceived as highly problematic. Frank W. Preston, in his erudite and often humorous article "Diversity and Stability in the Biological World," called into question the entire concept of stability in ecological systems. Arguing that population explosions or declines can actually be beneficial for certain organisms in the longer perspective of evolutionary time, Preston stated ironically that in some cases "fluctuation in numbers is necessary to stability, while lack of fluctuation is a threat to stability" (Preston 1969: 7). The chosen time frame, Preston pointed out, is very important, and even then relative stability over time is best described as more "dynamic" than "static." "It follows also that, in discussing stability, we have to make it clear whether we are considering consistency of numbers over a few days, . . . a few weeks, months, or years; or decades, centuries, or millennia. There is nothing very static about the organic world: the equilibrium achieved is very dynamic—if it is achieved at all" (Preston 1969: 8). The dynamic equilibrium of ecosystems was an idea that contributors returned to frequently throughout the symposium.

The challenge of formally describing this new concept of "dynamic equilibrium" was enthusiastically taken up by the mathematicians at the conference. R. C. Lewontin, a mathematical biologist from the University of Chicago, presented a paper entitled "The Meaning of Stability," in which he led fellow conferees through various equations and proofs in an effort to characterize what he called "the dynamical space" of a physical system. Margalef, who had proposed his unifying principles six years earlier, contributed an article entitled "Diversity and Stability: A Practical Proposal and a Model of Interdependence." Like Lewontin, Margalef relied on mathematics to assist in exploring the elusive qualities of stability. Margalef's solution was to combine measurements of diversity and "persistence," linking the two to form a "general expression of interdependence between species and between species and environment" (Margalef 1969: 25). He then derived equations from his general formula to describe the productivity of an ecosystem and its organization in space. Both Lewontin and Margalef were looking to solve the problem raised by Preston: that is, the temporal ambiguities inherent in the concept of stability. Their proposals were largely theoretical, and it remained to be seen whether the equations could be consistently applied in the field.

The last session of the conference finally turned to the practical "objective" that Woodwell and Smith had mentioned in their preface, the implications of human impacts on ecological systems. The chair of this session was biologist Garrett Hardin, whose environmental activist tendencies were well known in the scientific world. Hardin was also chosen to present the "symposium lecture," a keynote speech entitled "Not Peace, but Ecology." Not surprisingly, he chose to use his time to rally his fellow biologists and ecologists to fight against human encroachment on the natural world. Citing the publication of Rachel Carson's *Silent Spring* in 1962 as one of the "most important dates in biology in the 20th century," Hardin used the example of the application of pesticides to show the potentially devastating impacts of humans (Hardin 1969: 151). Hardin wished to emphasize the connections between the diversity of living organisms: "The most basic idea of ecology is that of a 'system.' . . . The practical implication of an ecological system is just this: *we can never do merely one thing . . .* this was the major message of Rachel Carson's book" (Hardin 1969: 152). It was also the lesson, Hardin implied, that biologists needed to recognize when discussing the importance of diversity and stability in systems. As Margalef had noted earlier, diversity's relationship to process provided a "model of interdependence." Hardin's provocative paper ended with the proposal of a new "Declaration of Interdependence," one that would stand against the polluting tendencies of our technological processes, the increasing human population, and the "unthinking worship of the god Progress" (Hardin 1969: 160).

Certainly, Hardin's paper is an exception in a collection of more objective scientific articles. But the fact that environmental concerns were even considered in the context of a respected scientific symposium is indicative of the way that diversity and stability were perceived at the time. Reducing an ecological system's diversity had serious implications, which were more than just a matter for objective scientific inquiry. It seems that the scientists at Brookhaven were unable to come to any agreement on the definition and measurement of the relationship between diversity and stability. The recognition of its importance to conservation, however, received a boost from the extensive discussions, and Brookhaven certainly represented a high point in the general debate about the significance of diversity in ecological systems.

Diversity-Stability after Brookhaven

The expanded attention to diversity and stability at this time is illustrated by the contents of the many ecology textbooks published in the years following the Brookhaven symposium. Charles Southwick's *Ecology and the Quality of Our Environment* (1972) included such sections as "Trophic Structure and Ecosystem Stability," "Influence of Man on Ecosystem Complexity and Stability," and "Species Diversity." Charles Krebs's text *Ecology* (1972) contained a substantial section called "Stability," which included a discussion of diversity's relationship to "the ability of a system to bounce back from disturbances" (Krebs 1972: 543). Eric Pianka's book *Evolutionary Ecology* (1974) had consecutive sections entitled "Species Diversity" and "Community Stability." Robert Whittaker, in the 1975 edition of *Communities and Ecosystems,* focused mainly on the evolution and generation of species diversity, but he also discussed the phenomenon of fluctuating, simple arctic communities as opposed to stable complex tropical communities. In short, it is evident that diversity and its functional role in ecosystems had become regular topics both for professional ecologists and for students of ecology.

However, the scientific debate over the diversity-stability relationship took an interesting turn in the early 1970s. From 1971 to 1973, Robert M. May published a series of articles and a book entitled *Stability and Complexity in Model Ecosystems* (1973b), in which he used mathematical proofs to show that in some instances it was likely that higher levels of diversity in ecological communities may result in decreased stability. As May concluded in one of his later papers, "All in all, rich trophic complexity and a diversity of different kinds of interaction between species is not conducive to qualitative stability. Insofar as the theory of qualitative stability relates to the muddied complexity-stability question, it is to re-echo the theme . . . that, in general mathematical models, increased complexity tends to beget diminished stability" (May 1973a: 641). It would seem that conservationists wishing to argue for maintaining diversity would find May's work disheartening. With no clear connection between diversity (or "complexity") and stability, the idea that diversity helped preserve a natural balance would hold little currency. In fact, the general idea of stability in the natural world was itself coming under fire in scientific circles, as hinted at in the Brookhaven papers that preferred to speak of the "dynamic equilibrium" of ecosystems.

But the idea that complex ecosystems were no more stable than simple ones played into a new environmental cause that was quickly gaining attention in the 1970s: the protection of tropical rainforests. For example, in one textbook entitled *Fragile Ecosystems: Evaluation of Research and Applications in the Neotropics* (1973), editors Edward Farnsworth and Frank Golley suggested that tropical ecological systems were highly susceptible to human-generated disturbances and had little inherent resilience for recovery. This new perspective was originally brought to the attention of the scientific world by Arturo Gomez-Pompa and his colleagues, in a 1972 article, "The Tropical Rain Forest: A Nonrenewable Resource." As the authors wrote, "It is the purpose of this article to provide a new argument that we think is of the utmost importance: the incapacity of the rain forest throughout most of its extent to regenerate under present land-use practices" (Gomez-Pompa, Vazquez-Yanes, and Guevara 1972: 762). Interestingly, instead of preserving diversity in order to maintain stability, this "new argument" essentially reversed the components: it was now important to maintain stability in order to preserve diversity. The authors were concerned with "the loss of millions and millions of years of evolution, not only of plant and animal species, but also of the most complex biotic communities in the world." In addition, they warned that "massive action [must] be taken to preserve this gigantic pool of germplasm" (Gomez-Pompa, Vazquez-Yanes, and Guevara 1972: 765). In short, highly diverse ecosystems were just as fragile as natural areas of low diversity, and it was this great variety that deserved protection. Diversity—of genes, species, and ecosystems—as suggested by Gomez-Pompa and his colleagues, had great value in and of itself. It was not essential to identify diversity as the underlying cause of stability. The implication was that it was more important simply to recognize that stability helped maintain diversity.

Still, the theory that diversity makes ecosystems more stable was firmly established in the mind of the environmentalist. By and large, after May and others had poked holes in the hypothesis, scientists lost interest in the search for the exact relationship between diversity and stability. One other conference, the First International Congress of Ecology (1974), explored further nuances of the diversity-stability hypothesis, but as with the conferees at Brookhaven, there was "considerable variation in attitudes regarding what ecosystem properties diversity is supposed to stabilize" (McNaughton 1977: 515). But conservation-minded scientists later voiced criticism that such uncertainty was because of the lack of empirical tests of the relationship. The reliance on theoretical work through mathematical models to disprove the hypothesis was seen as disconnected from real-world experience. In 1977, S. J. McNaughton reviewed numerous studies in his article "Diversity and Stability in Ecological Communities: A Comment on the Role of Empiricism in Ecology," and observed (with a humorous play on words): "I think the marked instability of attitudes regarding the diversity-stability relationship in ecosystems arises primarily from a low diversity of empirical tests of the hypothesis" (McNaughton 1977: 523). McNaughton concluded that the general hypothesis was much simpler than recent studies had suggested. At some levels in ecosystem processes (particularly the primary-producer level), the hypothesis is repeatedly shown to be true. While not an immutable ecological law that could be applied across all ecological functions, in general, McNaughton claimed, "species diversity mediates community functional stability through compensating interactions to environmental fluctuations among co-occurring species" (McNaughton 1977: 523). The "more elegant" and "more rigorous" mathematical models—as McNaughton somewhat derisively called them—were only able to show that there existed situations in which the diversity-stability theory would not hold true. But, as McNaughton argued, such models were inadequate in producing "reliable generalizations" of when it was apparent (through observation) that diversity did indeed mediate stability.

The debate largely disappeared from the literature in the later 1970s, but conservationists had come away from the confusion stronger than ever. Not only could they claim that diversity helped to keep ecological systems stable and functioning, they could also assert that diverse ecosystems were still highly susceptible to perturbation by human activity. Diversity, it seemed, did contribute to stability at certain trophic levels, but it did not necessarily make ecosystems more resilient. The inability of the scientific community to agree on any single theory did not seem to dampen the conservation community's enthusiasm for the topic. Arguably, this was because the idea of diversity, in and of itself, was growing in popularity as a sufficient motive for conserving natural areas.

Diversity in Conservation: Future Environments of North America

One important example of the conservation concern for diversity that arose out of the diversity-stability discussions of the 1960s was an influential conference that took place in 1965 under the sponsorship of the Conservation Foundation in Washington, DC. The conference organizers modeled the meeting on a successful 1955 Wenner-Gren Foundation conference, Man's Role in Changing the Face of the Earth, but planned a slightly smaller meeting with a narrower agenda. They were interested only in North America and were focused chiefly on what those interested in environmental issues could expect in upcoming years. Appropriately entitled Future Environments of North America, the conference invited "over forty scholars of many disciplines," including noted conservationists A. Starker Leopold, Ian Mc-Taggert Cowan, Samuel H. Ordway, Jr., Raymond Dasmann, Fairfield Osborn, Ian McHarg, and F. Fraser Darling, who served as chairman of the conference. The proceedings of the conference (1966) included not only the thirty-three commissioned papers, but also introductory comments, closing statements, and transcripts of discussions that took place after the paper presentations, revealing lively and articulate exchanges among the prestigious conferees.

John Milton, deputy director of international programs at the Conservation Foundation, provided a revealing "retrospect," in which he reviewed the successes of the conference's multidisciplinary approach to environmental issues. He said that the importance of "diversity" and "variety" had been a central theme of the meeting: "One of the significant truths to emerge from this conference was the value of diversity. Perhaps the most appalling aspect of modern man's insensitive degradation of the environment has been the mounting destruction of earth's natural diversity and the creation of monotonous, uniform human habitats. Ecology has shown us that varied ecosystems are healthy, relatively stable environments better able to withstand stresses; seen in this context, the contemporary trend toward creating an artificial, bland, standardized biosphere is a fundamental threat to the quality of human existence, if not man's very survival. With each loss of variety, our potential for human choice, freedom and change narrows" (Milton 1966: xvi). Milton not only was repeating the concerns that Elton had voiced in 1958 about the dangerous homogeneity of human development, he was also noting that "varied ecosystems" are more stable and resilient than the uniformity humans encourage.

These two themes appeared throughout the proceedings. Darling, in his introductory comments that opened the conference, mentioned that Charles Elton had been invited but was unable to attend. He commended Elton's call for "the preservation of variety and . . . diversification in the landscape." As Darling highlighted, "He [Elton] shows quite apart from any esthetic satisfaction in the varied scene, there is definite ecological advantage to be gained from diversity. There is a stability in such landscapes somewhat comparable with the ecological qualities of wilderness" (Darling 1966: 3). The influence of Elton's ideas on the planned objectives of the conference is evident.

Ian Cowan, in introducing the panel on "The Organic World and Its Environment," identified one principle that "underlies all discussions of biological conservation. This is that the maintenance of the best possible environment for future generations of men includes the maintenance of the entire gamut of variety in the living biota" (Cowan 1966a: 12). Cowan went on to give four reasons for choosing this principle. The first was the concern for "the world's complement of DNA in all its complexity" and the yet-undiscovered economic benefits that intact gene pools may provide. Second, Cowan cited the need to understand ecological processes and the important roles that seemingly unimportant organisms may play in ecosystems. Third, he perceived it necessary "to maintain tenuous and fragile species," claiming that "we shall gain more than we pay in attempting to cope with these very interesting problems [of endangered species]." Cowan finished by declaring that "there are larger moral issues separated from mere self-interest that suggest caution before we decide to commit to oblivion any single unique product of evolution on this continent" (Cowan 1966a: 12–13). In these four points, Cowan identified concern for genes, species, and ecosystems, as well as an intrinsic ethical value in maintaining the diversity of the natural world. Later, in his paper "Management, Response, and Variety," Cowan summarized his introductory statement: "The objective today . . . is directed toward the provision and maintenance of the maximum of variety. Each living organism is the repository of a unique assortment of biological information gained through the eras via the interplay of opportunity and response. Each offers a potential enrichment of human knowledge, experience, and enjoyment that is limited only by our capacity to appreciate. This capacity will certainly expand to levels yet unimagined and the loss of any single element in the world's store of varied life is an erosion of our potential" (Cowan 1966b: 56).

Other papers and the transcripts of discussions reveal that most conferees were supportive of the goal of maximizing diversity. One economist, Ayers Brinser, was in the uncomfortable position of dissenting, declaring that "diversity" was too general a concept to be useful in decision making. "I still believe that variety itself is no criterion for making decisions about how to proceed; that diversity does involve the matter of making choices. The really important thing is, how do you choose variety?" (quoted in Darling and Milton 1966: 219). But most others felt that the value of diversity was self-evident. F. Raymond Fosburg, who specialized in the restoration of degraded habitats, asserted that "variety gives us more flexibility, more self-correcting properties in the system" (quoted in Darling and Milton 1966: 303). Historian Clarence Glacken noted that plenitude—that is, beauty, variety, and complexity—was a major theme in the history of "man's attitude toward nature," one that defined an important human perception of the natural world. Finally, in the closing statement for the conference, Lewis Mumford reaffirmed the goal of maintaining diversity: "The real purpose of a conference like this—is it not?—is to insure the existence or the replenishment of a sufficiently varied environment to sustain all life, including human life, and thus to widen the ground for man's further conscious development. That ultimately, it seem to me, is what this whole business is about" (Mumford 1966: 722). Mumford, in his classic humanist language, linked the importance of natural diversity to the growth of the collective human mind: "Man needs the whole cosmos to sustain him. The knowledge of this cosmos and every living part of it enriches him, enables him to know himself for the first time" (in Darling and Milton 1966: 729). Even in this expanded perspective, variety and diversity remained the common themes.

The Future Environments of North America conference represents a telling snapshot of how leading conservationists were approaching and perceiving the environmental issues of the time. With diversity as a central theme, and with the conferees' expressed concern for gene pools, endangered species, ecosystems and habitats, and ecological process, the conference represents one of the first single events that brought together all of the elements of what would become the conservation concept of biological diversity. Throughout the 1970s, however, more advances would be made in tying all these interests together. In particular, the words and objectives of the Nature Conservancy during this time in the United States provide a representative example of how diversity gradually worked its way into the conservation vocabulary.

The Nature Conservancy and Protecting Land for Diversity

Throughout the 1960s, nongovernmental organizations representing a range of conservation interests proliferated. Many of these environmental groups had specific objectives (often associated with wildlife), but few dealt with protecting land as single-mindedly as the Nature Conservancy (TNC). TNC enjoyed the advantage of being one of the oldest and most respected environmental groups, and its method of purchasing land or otherwise negotiating protective strategies was well received by moderate conservationists who did not always appreciate the more strident voices of the newer, activist-minded environmental groups. Most important, TNC's contribution to the conservation dialogue and climate was significant to the recognition of diversity as an essential conservation concept.

An interesting example of TNC's evolving tendency to talk about diversity as opposed to land or nature is the numerous changes in their mission statement from 1972 to 1978. Up to 1972, the mission read, "The Nature Conservancy is the organization rallying the skills, the techniques, and the funds actually necessary to save land" (TNC 1972a: i). Later that year, the mission was changed to read more specifically that TNC "is the only national conservation organization . . .

whose resources are devoted to preservation of ecologically and environmentally significant land" (TNC 1972b: i). Four years later, the mission was again modified, identifying that TNC's "objective" was "to preserve and protect ecologically and environmentally significant land and the diversity of life it supports" (TNC 1976: i). Finally, in 1978 TNC changed its mission once more to read that the organization was "committed to the preservation of natural diversity by protecting lands containing the best examples of all components of our natural world" (TNC 1978: i). Gradually, TNC had shifted its emphasis from simply "land" to the diversity of elements that define the land. The increase in emphasis on diversity is significant, and the concept of natural diversity was a very important precursor to that of biological diversity.

During these years of the mission statement's evolution, a series of articles on the theory and practice of natural areas protection appeared in TNC's magazine, the *Nature Conservancy News,* several of which were written by Robert E. Jenkins, the director of science for TNC. Jenkins was a tireless advocate for the preservation of the complete range of ecosystems, and he seemed keen on clarifying that there were numerous purposes for protecting natural areas. He also identified the maintenance of diversity as the overall objective of TNC's preservation efforts. As he wrote in a 1972 article entitled "What *Is* a Preserve?": "We must exercise great selectivity in the use of our resources to preserve the highest quality natural areas for scientific research, different sorts of education, and especially to act as preserves or sanctuaries for species other than man. . . . The last epitomizes the very essence of most Conservancy projects, which implicitly emphasize the preservation of natural diversity" (Jenkins 1972a: 17). In a 1973 article entitled "Why Save Land?" Jenkins even more forcefully asserted the primacy of diversity: "As a logical minimum objective in a program of land preservation, it makes sense to try to preserve adequate representative samples of every kind of natural system and of the habitats of every sort of biological organism. The preservation of diversity in an increasingly constricted natural environment is of tremendous importance" (Jenkins 1973: 17).

One of Jenkins's early programmatic goals was a complete natural areas inventory, one that built upon the work of the International Biological Programme and the Federal Committee on Research Natural Areas. In discussing the importance of an inventory, Jenkins declared that protecting representative ecosystems was only a means to a much more detailed end. "What we are dealing with in nature preservation are tracts of land set aside as living space for the full diversity of biological species and types, as well as their physical environment with its own innate properties. This diversity represents genetic and other information, and each datum in the system is a potential or actual contributor to human well-being and/or the health and stability of the biosphere upon which we all depend" (Jenkins, 1972b: 18).

It seems that much of Jenkins's vocabulary of diversity sank deeply into TNC's institutional language. One could not discuss protecting land without describing the other natural components that would benefit. In fact, Jenkins and TNC even employed the term biological diversity, although it remained undefined. As Jenkins wrote in 1974, "The Nature Conservancy, as the national organization most identified with and involved in ecosystem preservation, has an obligation to see that the protection of ecosystems, biological diversity, endangered species habitat, scientific and educational areas, and 'ecologically significant lands' generally are not passed over as planning proceeds" (Jenkins 1974: 21). This early use of the term could have referred specifically to genetic diversity, or more generally, as Norse and McManus's definition would indicate, to different levels of diversity present in the natural world. Although we cannot know for sure, since no definition is offered, Jenkins's inclusive approach to conservation suggests the latter. Indeed, TNC adopted the term biological diversity to play a prominent role in the 1974 amendments to article 2 of their constitution, which listed the objectives of the organization. As amended, the first objective of TNC was "to preserve natural areas for biological diversity, for uses of science, and for wilderness experience" (TNC 1974: 16). It seems likely that the broadest possible meaning of the term was intended for use in such a general statement of purpose.

What Kind of Diversity? Natural, Ecological, Biotic

It is apparent, however, that *biological diversity* was soon dropped in favor of the term *natural diversity*. TNC released its report *Preservation of Natural Diversity* in 1975, and from that point on, *natural diversity* (or simply *diversity*) served as the most common reference in TNC articles and documents discussing the components of ecosystems. TNC's

concept of natural diversity had significant similarities to later definitions of biological diversity. But as presented in the report, natural diversity was focused chiefly on protecting land and only secondarily on
preserving species and genes. In laying out a conservation strategy,
TNC declared, "We need to set aside, in viable units, adequate examples of the full array of extant ecosystems, biological communities, endangered species habitats, and endangered physico-chemical environmental features. Only in this way can we maintain the full diversity of
genetic variability, ecological relationships, and special processes and
elements" (TNC 1975b: 10).

Understandably, in this report TNC defined natural diversity to ac-
commodate their traditional objective of preserving land. In a section
entitled "Why Save Natural Diversity?" TNC chose to discuss its reasons under three broad categories: the resource value of genetic diversity, the stabilizing value of ecological diversity, and the general scientific value of maintaining representative areas. The division between
genetic and ecological diversity matches Norse and McManus's original definition of biological diversity, and the discussion touches upon
many of the same concerns that Norse and McManus detail in the 1980
CEQ *Annual Report*. But it is apparent in TNC's language that natural
diversity was a synonym for a variety of large ecological entities that in
turn served to harbor species and genetic diversity. Biological diversity,
in its later usage, would be more inclusive in that it encompassed all of
the multiple levels.

One possible reason that TNC chose *natural* over *biological* as a
modifier for *diversity* was the introduction of TNC's State Natural
Heritage Programs, the objective of which "to provide the states with
the technical skills and methodology needed to develop a comprehensive program for the identification and preservation of natural elements of ecological significance" (Moyseenko 1974: 19). It also seems
likely that natural diversity was a term more easily understood and
more inclusive of nonliving natural features than the esoteric-sounding
term biological diversity. For whatever reasons, it was obvious that
TNC decided the word *natural* was an important descriptor for their
goals and objectives. Jenkins, in 1975 and 1976 articles about the Heritage Programs, chose to use both *natural* and *ecological* to modify *diversity* in his discussion of the importance of protecting ecosystems.

A crucial concept to the Heritage Programs, diversity was to become the main criterion for identifying whether states had protected an adequate sample of the full array of ecosystems. Part of the extensive inventory process that TNC was undertaking included classification systems for the "elements of diversity" present within each state (Wieting 1976: 31). With the help of newly designed computerized data-management programs, TNC wanted to ensure that the maximum amount of diversity was protected across the nation. Jenkins, in a 1978 article entitled "Heritage Classification: The Elements of Ecological Diversity," explained that one important way to select appropriate natural areas for protection was to remember that diversity occurred on several different levels. "The purpose is to organize what amounts to a dynamic atlas and data base of information on the existence, characteristics, numbers, condition, status, location, and distribution of occurrences of the elements of natural ecological diversity. From such a data base we can finally select our land conservation objectives with rigor and conviction" (Jenkins 1978: 30). This classification system evolved into what later TNC publications would call "*ecosystematics*— the systematics of ecological diversity*" (Radford et al. 1981: xvi). But such efforts seemed only to contribute to the proliferation of terms used in the business of nature protection. It was apparent that protecting ecosystems involved far more than simply setting aside areas of land. *Diversity,* with any number of modifiers, seemed the most appropriate way to communicate the multiple, complex objectives of modern conservation efforts.

While TNC was grappling with the idea of natural diversity, the term *biotic diversity* was also being used. In contrast to *natural diversity, biotic diversity* seemed to be employed by those concerned with species extinctions. In a 1974 conference sponsored by the Smithsonian Institution and the World Wildlife Fund, scientists and conservationists discussed issues of "biotic impoverishment." Species-saving efforts, particularly those of the WWF, were reviewed, and participants agreed that the traditional "species-by-species" approach was insufficient. As a result of this conclusion, "a major theme of the conference was the need to develop an 'ethic of biotic diversity,' in which such diversity is perceived as a value in itself and is tied in with the survival and fitness of the human race" (Holden 1974: 646). The conferees' main argument for preserving biotic diversity was that known and unknown benefits to humanity accompany a full complement of the world's species. In addition, while they noted the public relations value of "furry creatures with warm brown eyes," the focus turned to less well-known (and some unknown) species, ones that could become extinct before their true value was revealed: "Conservationists are well aware that the real problem is the salvation of countless other species, some known and some not, the silent majority, as it were, upon whose continued survival the quality of human life depends" (Holden 1974: 647). This type of dramatic language fueled the urgency of the situation and subsequently tied the concept of protecting diversity to an imperative that demanded quick and complete action.

Identifying Diversity as Valuable: TNC and the Natural Diversity Act

Whether described as natural, biotic, ecological, or biological, diversity was becoming the one characteristic of the natural world that seemed to best encompass all things of value for humans. In 1977, G. Jon Roush, the executive director of TNC, published an article entitled "Why Save Diversity?" in which he tried to summarize the major reasons for maintaining diverse natural environments. Roush listed four basic premises on which he believed "everyone should be able to agree, . . . any one of which makes the preservation of natural diversity an imperative of the most serious order" (Roush 1977: 9). The four premises were: "1. Diversity promotes the stability of ecosystems. 2. Diversity increases the possibility of future benefits. 3. Diversity is a source of human delight. 4. Protecting diversity is an ethical necessity" (Roush 1977: 9).

Essentially, Roush was describing various values of diversity, focusing on ecological, utilitarian, aesthetic, and ethical values. The ecological stability argument was set in the minds of environmentalists, no matter what scientific debate there was on the topic. The utilitarian criterion was largely based upon possible advances in agriculture and pharmaceuticals, which Roush illustrated with the wine industry's success in using genetic variety to improve the resistance of grape plants to disease, and with possible new cancer treatments derived from tropical plants. Roush characterized the capacity to enjoy the aesthetic value of diversity as "innate in every human being" and argued that the psychological benefits of the experience of a diverse natural world for a predominantly urban society were far more important than we could ever imagine. "Prolonged monotony of any sort produces neurosis, for which cultural and natural diversity are the only effective buffers" (Roush 1977: 12). Finally, Roush invoked Aldo Leopold in claiming that humans have an ethical obligation to tread on the land as lightly as possible, not only for the sake of maintaining options for future human generations but also for protecting other living things that have a right to exist. As Roush concluded, "We have inherited a world of extraordinary variety and complexity in which people have survived and flourished for eons. If we love that kind of world, we have no choice but to try to preserve its diversity" (Roush 1977: 12).

Roush's article serves as an excellent summary of the reasons why conservationists were enamored with the idea of diversity. Not only did diversity encompass all levels of the natural world that were worth protecting; it was also an effective concept for explaining the values that conservationists wanted to protect in the natural world. These values of diversity were so broad and inclusive that it was difficult to argue against efforts to preserve variety in nature. But the term diversity alone did not seem to have the distinction that would give it staying power in environmentalism. Roush sometimes employed TNC's preferred modifier *natural* in his discussions of diversity, but a strict definition was still lacking.

In the summer of 1977, a TNC-backed bill was introduced into the U.S. House of Representatives, proposing a "Natural Diversity Act." A companion bill was introduced in the Senate the next winter, and an article describing the Natural Diversity Act appeared in the January 1978 issue of the *Nature Conservancy News.* The authors of the bill were Congressman Keith Sebalius (R-Kansas) and Senator Lee Metcalf (D-Montana), but it was apparent that TNC staff had played a prominent role in drafting the language both of the act and of the journal article. The Natural Diversity Act was largely designed to take advantage of TNC's State Natural Heritage Programs and to provide funds "to systematically classify, inventory, monitor, and protect the nation's elements of diversity on a continuing basis" (Metcalf and Sebalius 1978: 10). States would receive substantial financial and technical assistance from a new federal office in the Department of the Interior and would cooperate with federal land management agencies to set criteria and standardize protection efforts.

Most significant, however, was a section in the journal article entitled "The Concept of Natural Diversity," in which the authors wrote:
"The term 'natural diversity' refers to the full array of biological
species—plants, mammals, birds, insects, reptiles, amphibians, fish,
snails, clams, and crustaceans—that have evolved over the last 600 million years, as well as to the different types of terrestrial and aquatic
communities and ecosystems into which these species are organized.
Collectively, these life forms constitute a vast genetic reservoir that
supplies the material for the continuing evolution of planetary life"
(Metcalf and Sebalius 1978: 7–8). Clearly, this description identifies all
three levels of the later concept of biological diversity: species, ecosystems, and the "genetic reservoir." However, the authors proceeded to
identify "tangible elements of diversity that can be located in the landscape" as targets of conservation efforts, and it is apparent that these
landscape features are designed to match TNC's longtime mission of
protecting land. The elements included terrestrial and aquatic community types, endangered species habitat, and "outstanding" geological
features. Certainly, the preservation of these entities were envisioned as
serving the purpose of preserving natural diversity, but the emphasis
on landscape-level goals was pervasive throughout the bill. Significantly, the authors of the bill also discussed "the value of natural diversity," in a section that listed the same four reasons that Roush had described months earlier, and the benefits are characterized as coming
from all levels of natural diversity, including the species and genetic
levels.

Thus, in many ways, the text of the Natural Diversity Act represents
an important precursor to the concept of biological diversity. It certainly demonstrates the significance that people were assigning to diversity and to its protection through land preservation. Unfortunately,
for whatever reasons, TNC's concept of natural diversity was not popular with environmentalists. The Natural Diversity Act itself never
made it out of committee, which is not surprising considering the extensive financial and administrative commitments it proposed. In addition, perhaps TNC's focus on landscape-level elements was not distinctive enough. The concept of natural diversity was simply lost in the
familiar goals of setting aside land. But the reasons for maintaining diversity that had been listed by Roush and by the Natural Diversity Act
no doubt carried over into the concept of biological diversity.

Conclusion

From the scientific debate on the relationship between diversity and stability, to the prominent references to diversity in the vocabulary of environmentalism, clearly the idea of protecting the inherent variety of the natural world had fastened itself in the conservation mentality by the 1970s. As a concept, diversity allowed people to encapsulate the concern for genes, species, ecosystems, and ecological processes. But while the unifying quality of diversity was evident, it was also apparent that the word *diversity* alone did not maintain the distinction that was needed to launch a campaign. The various modifiers—natural, ecological, biotic, biological—all were appropriate enough, but it seems likely that the interest and concern for diversity had not yet reached the necessary critical mass. By 1980 that situation apparently had changed. Perhaps because the idea of biological diversity was introduced at the right time into the right arenas, it could become established as the key umbrella concept in conservation. Certainly, however, the basis for the success of *biological diversity* was partially prepared by the broad discussion about the value of diversity itself in both scientific and environmental circles.

7 The Making of a Conservation Paradigm
A Confluence of Interests and Values

In the 1970s, the general interest in diversity, combined with the growing concern over the fate of the world's species, genetic resources, and ecosystems, represented a convergence of interests and values that lay the foundation for the rise of biological diversity as a leading conservation issue in the 1980s and 1990s. As previous chapters have discussed, there were several noteworthy early usages of the term biological diversity. In 1969, for example, N. W. Moore had written that the protection of biological diversity should be the "primary aim of conservation" (chapter 1). Five years later, the Nature Conservancy adopted the term as part of its organizational objectives (chapter 6). In addition to these appearances of the term, certain conservation interests were trying to link the protection of genes, species, and ecosystems to one programmatic goal. One result of this attempt to develop a single objective was the concept of the biosphere reserve (chapter 5), which was intended by UNESCO's Man and the Biosphere Programme not only to protect representative samples of the major ecosystems of the world but also "to conserve . . . the diversity and integrity of biotic communities of plants and animals," and "to safeguard the genetic diversity of species" (UNESCO 1974: 6). In short, different groups in the conservation community were testing out programs and ideas with the intention of trying to protect the full range of variety that humans value in the natural world, although no one had yet defined a single unifying concept that had taken hold successfully.

By the latter half of the 1970s, several important publications were serving as influential precursors to Norse and McManus's seminal

CEQ chapter. A number of these works used the term biological diversity in their text but did not provide any formal definition. Other publications did not use the term but provided the integrated perspective on conservation issues that Norse and McManus assumed in their definition of the term. But whether the actual term was used or not, it is evident that the idea of protecting the diversity of life on earth at multiple hierarchical levels was becoming more popular in the inner circles of the conservation community. Looking more closely at the ways in which environmental interests were approaching conservation in the late 1970s, it becomes apparent that Norse and McManus introduced the concept of biological diversity to a group that not only was receptive to the idea but had long been preparing an environmental network of interests to support such a cause.

This chapter provides a more detailed analysis of the ways in which the term biological diversity was used after the Norse-McManus definition in 1980. As reviewed in chapter 1, the concept played a prominent role in several key federal initiatives; in publications by well-known conservationists; and in the highly publicized National Forum on Biodiversity, which brought the issues surrounding the crisis to the attention of the scientific and media worlds. A number of other noteworthy publications and events also impacted the development of the concept, and their contributions are examined here. It is instructive to examine how interests with different agendas—a concern for genetic resource conservation, a primary focus on species protection, or perhaps an emphasis on a particular environmental value—employed the term to communicate their objectives. This all-inclusive aspect of the concept of biological diversity was certainly intended by those who introduced it, and because the concept was adopted by the full range of conservation interests, it moved quickly through the environmental community and soon caught the attention of national and international policy makers.

In a related trend, in the late 1970s and early 1980s the international environmental community was formulating policies and programs based on the idea that if the conservation of nature was to succeed in the developing world, the social and economic problems of poor countries needed to be addressed in concert with environmental issues. This philosophy became known as *sustainable development,* an approach to conservation that recognized the need to allow poverty-stricken nations to develop some of their natural resources in an intelligent, environmentally friendly way. Because biological diversity was represented as the "original resource," providing everything from food, clean air and water, materials, and medicines, it was immediately incorporated into the sustainable development dialogue. This linkage to economic well-being was emphasized by Norse and McManus, and it contributed significantly to the speed with which the concept of biological diversity spread.

Many environmentalists believed that such popularity came with a price. Because of biological diversity's elevation to a status rivaling that of more traditional social issues, some felt that the original environmental objectives in the concept would become weakened by manipulation and compromise. But others felt that if conservation issues were to advance in the public arena, the environmental community would have to make the case that protecting the integrity and health of the natural world was in every human being's best interest, and that environmental problems needed to be considered on the same level as other global issues such as poverty and social justice. Whether this popularity served to advance conservation efforts or to detract from them has remained a point of contention for many years.

Certainly, the debate continued on biological diversity's strengths and weaknesses as a broad, all-inclusive conservation concept. But it is difficult to argue against the observation that the concept's inclusiveness not only linked previously separate environmental interests but brought the environmental crisis to the attention of national and international policy communities. The most significant contribution of biological diversity as a conservation paradigm was to heighten awareness of several interconnected crises in the human use of the natural world, and to bring people together in an effort to find solutions to those crises. In accomplishing this, the concept filled a need that had been apparent in the conservation world for many years.

Precursors to the First Definition: Lovejoy and the National Research Council

While the first published definition did not appear until 1980, there were a small number of authors who employed the term without providing any formal description of what biological diversity meant.

These precursor publications reveal some important roots of the concept by highlighting the connection betwe en the future concern for biological diversity and past concerns for gen tic, species, and ecosystem diversity. One of the hallmarks of Norse and McManus's concept was to consciously combine concerns for different hierarchical levels under one conservation objective. But as noted in chapter 1, the first definition of biological diversity contained only two divisional levels: genetic and "ecological" diversity, the latter being a synonym for what the authors called "species richness." It is interesting that two of the most often-cited usages of the term before the 1980 definition represented these two levels. The three-tiered definition would appear later.

An early prominent defender of maintaining species diversity was Thomas Lovejoy, a noted author, scientist, and conservationist whose name has since become closely linked with efforts to protect biodiversity. Lovejoy, who has held several influential positions in the conservation world, including posts with the Smithsonian Institution and the World Wildlife Fund, is often given credit for first using the term biological diversity. While this does not seem to be accurate historically, there is little doubt that Lovejoy has played an important role in popularizing and educating others about the issues surrounding the biodiversity crisis since the mid-1970s.

Lovejoy identified biological diversity as a significant future environmental issue in two different works published in 1980. First, in the foreword to *Conservation Biology,* the collection of papers from the 1978 International Conference on Conservation Biology edited by Michael Soulé and Bruce Wilcox, Lovejoy declared that the "reduction in the biological diversity of the planet is the most basic issue of our time" (Lovejoy 1980b: ix). However, while Lovejoy's personal concern for the degradation of the living world likely included a concern for all levels of diversity, it is apparent that in this particular context, Lovejoy was primarily focusing on the threat to global *species* diversity. In the first paragraph, he wrote dramatically about the plight of tropical species, referring to a contemporary study "that predicted the loss of two-thirds of all tropical forests by the turn of the century. Hundreds of thousands of species will perish, and this reduction of 10 to 20 percent of the earth's biota will occur in about half a human life span" (Lovejoy 1980b: ix). The emphasis on species diversity was appropriate, for the articles in the book were dedicated to the conservation of species, and all discussion of ecosystem or genetic diversity was presented in this context. Lovejoy, in this short piece, apparently did not have the same goal as Norse and McManus in introducing a new concept for unifying the conservation world; rather, he focused on a more specific issue.

This observation is also true for the other noteworthy 1980 publication by Lovejoy, a subsection in the *Global 2000 Report to the President,* entitled "Changes in Biological Diversity." Lovejoy's piece was part of a larger section called "Global-Scale Environmental Impacts," located in a chapter entitled "Forestry Projections." Interestingly, Lovejoy at this time was working in a nearby building to Norse and McManus in Washington, DC, but the writers never conferred on the topic or their word choice during the period in which they were researching and writing (Lovejoy 1999; Norse 1999). Again, even though Lovejoy chose to highlight *biological diversity* in the section title, he offered no recognizable definition. In fact, the title is the only place he used the term, although he did provide several near synonyms throughout the text: "biotic diversity," "diversity of flora and fauna," "biological resources," "biotic resources and their contributions," and "biological capital." While some of these additional terms might suggest the idea of genetic diversity that Norse and McManus emphasized as one of the two levels of biological diversity, Lovejoy again chose to focus on species diversity and the projected losses of species in the near future. In fact, a major part of the article was devoted to Lovejoy's species-area curves, which predict the percentage of species lost per area of tropical forest cleared—a calculation he based on the theory of island biogeography pioneered by MacArthur and Wilson in the 1960s. He characterized the flora and fauna of the planet as "the only truly nonrenewable resource . . . when extinct biotic resources and their contributions are lost forever" (Lovejoy 1980a: 327). While Lovejoy may not have provided a glossary definition, and while he was clearly concerned with species over other types of diversity, he was still expressing primarily the same concern for "living resources" that Norse and McManus expressed in their defining article. Most important, his use of the term biological diversity in this context implicitly connected it to the tradition of species protection efforts in Western conservation.

While Lovejoy's focus was on protecting species, an even earlier appearance of the term biological diversity focused on the issues surrounding the importance of protecting diversity at the genetic level. This frequently cited document, a report called *Conservation of Germplasm Resources: An Imperative,* was published in 1978 by the National Research Council (NRC). As in Lovejoy's work in *Global 2000, biological diversity* appeared here as part of a section title—"Natural Ecosystems and Biological Diversity"—and was not defined in the text. Apparently it was considered, as in Lovejoy's works, a general descriptor not needing further clarification. But, as noted, while Lovejoy viewed the conservation of biological diversity through the lens of species preservation, it was clear that the NRC perceived it as a function of their concerns surrounding the maintenance of genetic variability and the protection of germplasm resources: "Saving the rich diversity of genetic material that has been provided by natural mutation and evolution can be achieved and is worth whatever effort may be required" (NRC 1978: 4). The authors did recognize, however, that preserving genetic diversity meant preserving both species and ecosystems, and they focused in particular on the importance of protecting natural habitats as the most "reliable method" for conserving species and genetic variability: "It seems obvious that an adequate number of . . . ecosystems must be identified and . . . protected if the world's germplasm resources are to be preserved" (NRC 1978: 9). In addition, later chapters focused on the preservation of economically important plants, animals, and microorganisms, and reviewed methods of collecting for research and teaching as well as the status of cryobiological preservation (freezing germplasm for future use). From the subjects addressed and the sources cited, it is evident that the authors of the NRC report were drawing heavily upon the publications and conferences of the 1960s and early 1970s that defined the genetic resource conservation movement. In this way, the initial link was made between the term biological diversity and earlier efforts to protect genetic variability.

Even though the NRC report title suggests a narrow concern, its broad approach to conserving "germplasm resources" makes it a notable predecessor to the defining CEQ chapter. The NRC report was concerned with all kinds of germplasm, not just economically viable genetic material, and it noted the potentially harmful impact that reducing overall diversity can have on other entities in ecological systems. "A natural ecosystem with a diminished diversity of living systems is . . . an impoverished system and any organism in a natural ecosystem with a contracting genetic diversity is a threatened organism. It is in the best interest of human society to see that the diversity of natural ecosystems does not appreciably diminish" (NRC 1978: 5). But the authors of the report failed to place their concern for "living systems" within a well-defined conceptual framework. Not only was there no definition offered for biological diversity, but the term was used too infrequently for readers to extract a definitive meaning from the text. *Germplasm resources* was perhaps too esoteric a term to become popular as a conceptual focus for conservation efforts. Although the NRC defined germplasm resources as "the total array of living species, subspecies, genetically defined stocks, genetic variants, and mutants whose continuing availability is important for society's present and future health and welfare" (NRC 1978: 1), the concept of "germplasm" seemingly suggested a less inclusive perspective. Still, by highlighting biological diversity in the title of a chapter, the NRC took a first step in tying the term to the broad conservation issues surrounding the protection of genetic resources. Together with Lovejoy's works, the simple appearance of the term in association with the concern over species depletion and the loss of genetic diversity established a link that would be strengthened in future uses of the concept.

Erik Eckholm's *Disappearing Species*

One significant document that is neglected in the few publications that discuss the history of the concept is Erik Eckholm's *Disappearing Species: The Social Challenge* (1978). Eckholm was a senior researcher with the Worldwatch Institute, and his thirty-eight-page work was published in a series of short monographs, Worldwatch Papers, dedicated to environmental issues. The volume was widely acclaimed when it came out, viewed by many as one of the first works to move beyond concerns limited to the popular endangered megafauna species. As Eckholm observed, "At risk, the scientists say, are not just hundreds of familiar and appealing birds and mammals. Examination of the survival prospects of all forms of plant and animal life—including obscure ferns, shrubs, insects, and mollusks as well as elephants and wolves—indicates that huge numbers of them have little future. Not hundreds, but hundreds of thousands of unique, irreplaceable life forms may vanish by the century's end" (Eckholm 1978: 5).

While the term biological diversity is not highlighted in the title, Eckholm used it (without a definition) in several passages, far more than either Lovejoy or the NRC authors. In addition, while Eckholm was concerned chiefly with species (as his title and the above passage suggest), his discussion reached far more broadly than the other studies, touching on a range of issues similar to the ones raised in the CEQ report. For example, in a section entitled "Biological Impoverishment: The Human Costs," Eckholm wrote, "Probably the most immediate threat to human welfare posed by the loss of biological diversity arises from the shrinkage of the plant gene pools available to agricultural scientists and farmers—a critical if largely separable, aspect of the more general problem" (Eckholm 1978: 12). He continued for several pages about agricultural issues and the potential losses of economic values and then turned to issues of ecological function: "Beyond particular economic or scientific losses caused by the destruction of particular species lies a more basic threat: the disruption of ecosystems on which human well-being depends" (Eckholm 1978: 18). Eckholm even anticipated the importance of ecosystem diversity, the organizational level that Norse and McManus had opted against including in the CEQ chapter. Eckholm commented: "The overriding conservation need of the next few decades is the protection of as many varied habitats as possible—the preservation of a representative cross section of the world's ecosystems, especially those particularly rich in life forms" (Eckholm 1978: 27). Discussing more than disappearing species, this paper was about the overall loss of variety. *Biological diversity* was just one descriptor that Eckholm found useful, along with other terms such as *natural diversity, biological wealth, genetic heritage,* and *flora and fauna.* In short, Eckholm skillfully outlined the concept without formally providing it with an official name, instead allowing his information-rich discussions to provide their own framework.

Most important, as evident in the above quotations, Eckholm was consciously bringing together the work of different segments of the conservation community. From agricultural concerns about germplasm, to traditional environmental worries about species loss, to national and international efforts to protect ecosystems and ecological processes, Eckholm was the first to articulate the full range of problems facing conservationists in a concise, clear document. The broad vision with which he encapsulated the crises surrounding the living diversity of the natural world was remarkably prescient, and it significantly advanced the idea of the need for a unifying objective for all conservation.

Eckholm's paper was not only a significant precursor to the concept of biological diversity in terms of its representation of the different conservation traditions; it anticipated an important trend in linking conservation and development concerns, a trend that would come to characterize many political discussions involving biological diversity and sustainable development. First, Eckholm argued that in order to preserve the biological capital from which humans obtain their resources, economic development must proceed, especially in the poorest nations. Eckholm believed that the greatest pressure on the earth's living diversity would come from those people whose lack of fundamental goods and services was deplorable: "Their social deprivation will corrode the foundation of even the best-designed species-preservation structures" (Eckholm 1978: 21). According to Eckholm, a conservation plan could not be successful without the connection to human well-being. As his title indicates, he saw the problem as a "social challenge," in which managing the affairs of human systems mattered just as much as establishing protection for biological systems. "Clearly, the struggle to save species and unique ecosystems cannot be divorced from the broader struggle to achieve a social order in which the basic needs of all are met" (Eckholm 1978: 24).

Second, Eckholm believed that this connection to economic progress might serve as the bridge between developers and conservationists all over the world. To grow economically, developers needed a healthy natural resource base; to protect biological diversity, conservationists needed to ensure economic and social stability. In addition, this interchange was of paramount importance in light of the growing human population in the areas where people were poor and biological diversity was rich. As Eckholm concluded, "The eventual tripling in human numbers projected by many demographers would simply be incompatible with the preservation of needed natural diversity. Locally and internationally, economic orders must be created that are at once ecologically and socially sustainable. Developers and conservationists need each other if the ultimate goals of either are to be met, for biolog-

ical impoverishment and human impoverishment are inextricably intertwined" (Eckholm 1978: 32).

By presenting the problem as one that could bring sworn enemies
together, Eckholm was helping to lay the foundation of an all-encompassing concept for how humans should live in the world. Here was a
cause that all could fight for, because the protection of diversity arguably served all parties' interests. This emphasis on sustainability, inclusiveness, and interdependency makes Eckholm an obvious source
not only for the CEQ chapter, but also for several other defining publications and events in the early 1980s.

The Concern for Tropical Moist Forests

Another important precursor to the concept of biological diversity,
one that was intimately tied to the rise of the term in environmentalism's vocabulary, was the concern over the losses of tropical moist
forests. The plight of the "rainforests" (as they are more popularly
called) became well known to people outside of traditional conservation circles as a result of campaigns by environmental groups that often
used exotic endangered species to capture the public's attention. "Save
the rainforest" became a slogan in the 1980s and 1990s almost as ubiquitous as the earlier "Save the whales" campaign. Calls for tropical
rainforest conservation came as early as the 1960s, but it seems that (as
discussed in chapter 6) a 1972 article in *Science,* entitled "The Tropical
Rain Forest: A Nonrenewable Resource," served as the catalyst that
spurred popular conservation and scientific interest. The authors, A.
Gomez-Pompa, C. Vazquez-Yanes, and S. Guevara, were faculty members in the department of botany at the Institute of Biology, National
University of Mexico. The purpose of their article, they declared, was
"to provide a new argument that we think is of utmost importance: the
incapacity of the rain forest throughout most of its extent to regenerate
under present land-use practices" (Gomez-Pompa, Vazquez-Yanes,
and Guevara 1972: 762). Humans were irreversibly changing the tropical ecosystems of the world. The implications for the species of these
forests were inescapable—the conclusion, quite simply, was the "great
danger of mass extinction" (Gomez-Pompa, Vazquez-Yanes, and Guevara 1972: 764). Although the scientific evidence to prove these claims

was apparently still incomplete, the authors felt that the possible consequences of waiting "for a generation to provide abundant evidence" were not acceptable—by then the rainforests' ecosystems might be destroyed. As Gomez-Pompa and his colleagues concluded, "The sole fact that thousands of species will disappear before any aspect of their biology has been investigated is frightening. This would mean the loss of millions and millions of years of evolution, not only of plant and animal species, but also of the most complex biotic communities in the world. We urgently suggest that, internationally, massive action be taken to preserve this gigantic pool of germplasm by establishment of biological gene pool reserves from the different tropical rain forest environments of the world" (Gomez-Pompa, Vazquez-Yanes, and Guevara 1972: 765). In this final passage, the authors hit upon all three hierarchical levels that would come to characterize the concept of biological diversity: "plant and animal *species,*" "biotic communities" (*ecosystems*), and "germplasm" (*genes*). Concern for the rainforests was ultimately based on the concern for their rich biological diversity. The concept simply had yet to be named and formally defined.

Interest in tropical conservation grew throughout the 1970s, as evidenced by an increasing number of articles, books, and conferences on the topic. But an influential and comprehensive characterization of the rainforest crisis was not provided until Norman Myers published his extensive research in the late 1970s and early 1980s. Myers's ostensible goal was to provide the "abundant evidence" missing from the original study by Gomez-Pompa and his colleagues. Myers compiled data collected in personal surveys and from previous studies conducted in tropical moist forests all over the world. His technical work was published in 1980 as a report for the Committee on Research Priorities in Tropical Biology, of the National Research Council, which had published the study on germplasm resources in 1978. The report, entitled *Conversion of Tropical Moist Forests,* received considerable attention. In it, Myers provided a brief introductory section on "biotic diversity," warning that many of the tropical moist forest species (mostly insects) were "unusually susceptible to extinction" because of their "highly specialized ecological requirements; . . . low densities; . . . and localized distributions. . . . Elimination of a substantial proportion of the planetary spectrum of species will mean a gross reduction of life's diversity on earth; it will entail a permanent shift in the course of evolution, and an irreversible loss of economic opportunity as well" (Myers 1980a: 14).

The theme of lost economic opportunity is the topic for which Myers ultimately became best known. His most influential work, arguably, was not the NRC report but a book, published the previous year, directed at more popular audiences: *The Sinking Ark: A New Look at the Problem of Disappearing Species* (1979). Myers's "new look" was to perceive the extinction crisis as a crisis of resources, a potential loss of immense utilitarian value: "There can hardly be a more valuable stock of natural resources on earth than the planetary spectrum of species with their genetic reservoirs" (Myers 1979: x). It should be noted that Myers did not use the term biological diversity anywhere in the book. But there were many significant similarities between *The Sinking Ark* and Norse and McManus's definitional chapter in 1980. For example, the three major categories that Myers listed as his "utilitarian" arguments for preserving species paralleled the topics comprising the initial section of the CEQ chapter on "Material Benefits of Biological Diversity": Myers focused on benefits to "agriculture," "medicine and pharmaceuticals," and "industrial processes" (including energy production), while Norse and McManus wrote sections on biological diversity as a source of "food," "medicines," "industrial chemicals and raw materials," and "energy." Myers's influence is undeniable; not only did Norse confirm that Myers's work on tropical forests spurred the CEQ to propose the chapter on "the status of life on Earth," but as noted in chapter 1, Norse and McManus cited *The Sinking Ark* nine times.

Yet Norse and McManus diverged from Myers by expanding their defense of preservation beyond the utilitarian benefits. While Myers did mention the existence of aesthetic and ethical values in conserving species, his aesthetic argument consisted mostly of trying to assess how much beauty is worth to people (through how much they pay for certain aesthetic amenities), and he discounted the ethical argument as filled with inconsistencies and impracticalities. Norse and McManus, in contrast, opted to compose a section on the "psychological and philosophical basis for preserving biological diversity," which included discussions of aesthetic, ethical, recreational, and humanistic values. Although not as long as the section on material benefits, it is evident that the authors felt that such arguments appealed to significant numbers of people and were therefore worth including. Myers's firm position was utilitarian. Norse and McManus—appropriately enough—seemed to favor having a diversity of arguments in their arsenal.

But the impact of Myers's concentrated research on the economic value threatened by species extinctions should not be discounted in any way. Before Myers, environmentalists often dreaded economic evaluations, for it seemed those who wished to develop the land could always prove their projects would generate more revenue than the land would be worth undeveloped. Myers's work shifted the advantage, at least in the tropics, to the conservationists. Species and their inherent genetic diversity have a potentially high monetary value, and Myers's utilitarian arguments not only provided the environmental community with a new and potent weapon but also captured the attention of policy makers who would not previously have cared about losing unique examples of biological diversity. For this reason, the utilitarian economic arguments for protecting nature's variety became an important and frequent component of the concept. While Norse and McManus were able to provide some foundation for other types of values, the characterization of biological diversity as first and foremost a source of great material wealth held primacy, particularly in the early formative years of the concept's evolution.

World Conservation Strategy: Protecting Species, Genes, and Ecosystems through Sustainable Development

Myers's emphasis on economic value, combined with Eckholm's observation that social and economic issues must be considered in any overall plan to protect biological diversity, presaged an international environmental movement that ultimately would serve to bind the concept of biological diversity with the idea of sustainable development. Throughout the late 1970s, while Myers, Eckholm, and others were publishing articles about tropical deforestation and species extinctions, a group of international organizations was working on successive drafts of a document intended "to represent a consensus of policy on conservation efforts in the context of world development" (IUCN

1980: ii). The final text was released in 1980, entitled *World Conservation Strategy: Living Resource Conservation for Sustainable Development.* The Strategy Conference was prepared by the International Union for the Conservation of Nature and Natural Resources (IUCN), with support from the United Nations Environment Programme (UNEP) and the World Wildlife Fund (WWF). While the term biological diversity was not used in this publication, the subject matter and the hierarchical arrangement of concerns were strikingly similar to that which would come to characterize the movement behind biodiversity conservation.

First, the IUCN's focus on "living resource conservation" closely paralleled Norse and McManus's concerns. In their CEQ chapter title, "Ecology and Living Resources—Biological Diversity," "living resources" was presented as being at least partially synonymous with *biological diversity.* In their definition, Norse and McManus identified two related concepts, genetic and ecological diversity, as being essential components of ecological processes. Significantly, the World Conservation Strategy (WCS) highlighted three "main objectives of living resource conservation: a. to maintain essential ecological processes and life support systems. . . . ; b. to preserve genetic diversity (the range of genetic material found in the world's organisms . . . ; c. to ensure the sustainable utilization of species and ecosystems (notably fish and other wildlife, forests and grazing lands)" (IUCN 1980: vi).

While "diversity" is mentioned only in reference to preserving genetic resources, it is evident that the IUCN was considering an approach to conservation that identified the different levels of the components of living resources. From genes, to species, to ecosystems and "essential ecological processes and life support systems," the WCS proposed that conservation be approached hierarchically. The similarity to the CEQ chapter as well as to Eckholm's Worldwatch paper indicates an obvious commonality in thought in the conservation community around 1980.

It is evident, however, that the WCS had a use-based bias, identifying in particular species and ecosystems most valued by humans for their products. This quality alone makes the WCS's concept of "living resources" different from Norse and McManus's idea of biological diversity. But the way in which the WCS was drafted makes it an important document to consider. Because the WCS was "the product of an extremely thorough consultation process, inevitably reflect[ing] a compromise" (IUCN 1980: ii), it seems reasonable to assume that this hierarchical, all-encompassing approach to conservation (emphasizing importance of the three levels of genes, species, and ecosystems) was one that resonated with the diverse set of important players at that time. As noted in the WCS preface, "IUCN's membership currently consists of more than 450 government agencies and conservation organizations in over 100 countries. These members were first polled for their views on conservation priorities. Subsequently two drafts of the WCS were sent to them for comment, as well as to IUCN's Commissions of more than 700 scientists and other experts" (IUCN 1980: iii). If this is true, then while Norse and McManus listed only two levels in their 1980 publication, it seems likely that the environmental constituencies that were considering the same issues were already thinking about ecosystem maintenance and variety in conjunction with genetic and species diversity conservation. Thus, the three levels that would later be delineated in the definition of biological diversity were well represented in the dominant conservation interests of the times.

As mentioned, one notable characteristic of the WCS was its emphasis on sustainable development. Some conservationists likely viewed this component as one of the unfortunate compromises that arose in the drafting process. But in the idea of conserving living resources for the good of humanity, the WCS (like Eckholm) concluded that it would be unreasonable to assume that we can completely freeze development around the world, above all in poorer countries. This perspective had even earlier expression in the United Nations' environmental conference in Stockholm in 1972. A matter of urgency, the WCS notes, was that "hundreds of millions of rural people in developing countries, including 500 million malnourished and 800 million destitute, are compelled to destroy the resources necessary to free them from starvation and poverty" (IUCN 1980: vi). It was seen as a vicious cycle. The growing population exploited the environment in their effort simply to survive and provide for families; the resultant degradation of resources perpetuated the health, social, and economic problems; people trapped in poverty tended to have more children; and the process painfully started anew.

Thus, the IUCN perceived that an effective conservation strategy not only must provide a plan to protect genes, species, and ecosystems and ecological functions, it must also work to change the conditions that cause humanity to destroy the living resources on which they depend. Indeed, the first two obstacles to achieving conservation noted in the WCS were "*a. the belief that living resource conservation is a limited sector,* rather than a process that cuts across and must be considered by all sectors; *b. the consequent failure to integrate conservation and development*" (IUCN 1980: vi; authors' emphasis). This holistic approach was designed to bring together often opposing sides for attainment of a common good. As a wake-up call to the developers, the message from conservationists was familiar: living resources are essential in securing both raw materials and the social stability that are required for sustained economic growth. But there was also a message to the conservationists, emphasizing that they had to be willing to work with development so that it would be done intelligently and humanely. For some, on both sides, such cooperation was a kind of unholy marriage. But because the crises outlined in the WCS were conceived as imperatives that concerned everyone's interests, the publication garnered a wide audience in a cross-section of conservation groups, governments, and industry.

The 1981 Strategy Conference: Focusing on "Vital U.S. Interests"

The impact of the World Conservation Strategy approach may be seen in the first major event that addressed concerns of biological diversity loss by name. The U.S. Strategy Conference on Biological Diversity took place in November 1981 and was sponsored—as discussed in chapter 1—by a wide range of agencies and programs. This broad interest, especially from such agencies as the U.S. Agency for International Development (USAID) and the State and Commerce Departments, was generated at least in part by the release of the WCS, in addition to the NRC reports on germplasm resources and tropical deforestation in the previous year. As noted in the preface of the proceedings for the conference, "The World Conservation Strategy and various technical studies, such as those carried out by the U.S. National Academy of Sciences have highlighted the pressures which modern civilization imposes on life-forms and have called for accelerated international efforts to counter what they project to be increasingly

rapid losses of genetic resources" (USAID 1982: iii). The issues surrounding the crisis had been perceived as serious enough to warrant a
planning meeting of USAID and State Department officials in early
1981. Their concerns included the "proper role for the U.S. Government vis-à-vis . . . the protection of biological diversity," and the "vital
U.S. interests involved—economic, environmental, political, and strategic." The conference was planned "to promote public awareness of
the linkages that exist among diversity issues; to engage federal agencies in a dialogue with business, academic and scientific sectors, and
congressional representatives; and to develop specific recommendations for domestic and international responses" (USAID 1982: iii).

In the introductory statements of the State Department and USAID
officials, the influence of both the CEQ chapter and the WCS was
clear. Biological diversity was characterized as the "original" resource,
much in the manner of Norse and McManus. "[On] this threatened
biological diversity depends, in significant degree, the fundamental
support system for man and other living things," declared James Buckley, undersecretary of state for security assistance, science, and technology (USAID 1982: 9). The close connection between conservation
and development was also highlighted. Dr. Nyle Brady, senior assistant
administrator for science and technology for USAID, reminded conferees that "the prognosis for large numbers of threatened plant and
animals is closely tied to the fate of millions of impoverished people
whose own existence is threatened—threatened by lack of opportunity, wretched living conditions, high birth rates, and unemployment"
(USAID 1982: 12).

But the proceedings of the conference reveal that the U.S. government's timely interest in biological diversity was fueled significantly by
one subject in particular: the threat to genetic resources and the potential loss of opportunity in the agricultural and business sectors of the
global economy. In fact, it is somewhat unclear how much participants
distinguished between the "biological" and "genetic" diversity. Of the
thirteen principal conclusions, ten addressed problems surrounding
the loss of genetic diversity or germplasm resources. Only two mentioned "species" diversity, and contextually it is apparent that preserving species was chiefly considered a function of preserving their genetic material. The first speaker, Dr. William L. Brown, chairman of
Pioneer Hi-Bred International, began his talk by clarifying that his

"comments today will deal solely with biological diversity of cultivated plants" (USAID 1982: 13). This usage of the term was obviously narrower than the umbrella concept that Norse and McManus had proposed in their CEQ chapter. Titles of other papers presented at the conference included: "Biological Diversity: Basic for Agricultural Success"; "Genetic Diversity: Strategic Significance and U.S. Opportunity"; "Genetic Conservation and the Role of the United States"; and perhaps most revealing in terms of the newest utilitarian interest in diversity conservation, "Biological Diversity and Genetic Engineering." In contrast to these presentations, there were a few general treatments of the topic as a broader, unifying concept for conservation. David Pimentel of Cornell University discussed a wide range of benefits in his paper, "Biological Diversity and Environmental Quality." Thomas Lovejoy, in "Biological Diversity and Society," was particularly careful to note the contribution of biodiversity at all levels, from genetic resources to the "role the biota collectively play in the stability of global chemistry and climate" (USAID 1982: 50). But the participants concerned with germplasm for human use dominated the conference.

This imbalance was not lost on some of the conferees, a number of whom addressed concerns for other values they believed were being ignored. Dr. Carleton Ray, a marine biologist from the University of Virginia, commented that the ecological value of biological diversity was being overlooked: "I'm not quite certain how the subject of biological diversity has become translated into what we've heard most about today, namely preservation of genetic material and species. . . . In order to maintain biological diversity, one has to pay a lot of attention, particularly in aquatic systems, to sustainability of systems and to preservation of ecological processes" (USAID 1982: 57). Dr. Archie Carr of the New York Zoological Society pointed out that besides the material benefits of biodiversity, there were also humanistic and aesthetic values at stake: "The record as we have created it thus far today may be incomplete. . . . I would like to submit that one important reason for preserving species is because people like them. There is an aesthetic side to all this, a side that is admittedly difficult to work with, but one that is important not to ignore" (USAID 1982: 59). Dr. Michael Soulé, as the leading conservation biologist, took Carr's comments one step further, adding that certain ethical values needed to be recognized: "It is regrettable that we must all pretend to be concerned exclusively with man and his welfare and put nearly all of our arguments for conservation of biological diversity in terms of benefit for man. I think we will have reached cultural adolescence when we can admit in public that conservation is not only for people, something most of us admit already in private" (USAID 1982: 61).

But while arguments based on aesthetics, ethics, or the maintenance of ecological functions were forced to the back seat by the genetic resources issues, it seems likely that without the interest in germplasm conservation there never would have been a conference at this time focusing on biological diversity. As James Buckley observed in his introductory statement, "Recent well-publicized breakthroughs in genetic engineering may be what is required to focus public attention on the explicit interest each one of us has in seeing that the global stock of irreplaceable genetic material isn't squandered. This may help us focus attention on the practical need to conserve our biological resources" (USAID 1982: 10). By using *biological diversity* in the title of the conference, the State Department and USAID, although chiefly interested in the status of genetic resources, had allowed the opportunity for a much broader discussion to take place. Concerns for germplasm and genetic raw material for future biotechnological breakthroughs characterized the conference. But as Buckley's words imply, these "practical" matters could help to raise awareness of an issue that impacted a wide collection of interests. In addition, the global connections between the loss of biological diversity and the well-being of the economic, political, and social sectors were revealed and even highlighted. Most important, the term biological diversity had been officially introduced to the United States policy-making community, and although it was initially treated as synonymous with "genetic" variety, the conference provided an ample platform upon which a more broadly informed concept might have room to evolve. Biological diversity, as conceived by Norse and McManus, had gained its first real foothold—not only in national conservation circles but on the larger stage of national policy consideration as well.

"Natural" Diversity and the NFMA Diversity Clause

About the same time that the use of the term biological diversity was rising in popularity, there was another term being used mostly by land managers and resource professionals: *natural diversity.* Natural diversity had been a concept used by the Nature Conservancy, first in a 1975 study entitled *The Preservation of Natural Diversity,* and even more prominently in a TNC-backed bill first introduced in Congress in 1977, the Natural Diversity Act (see chapter 6). Although primarily focused on landscape-level conservation, the definition of natural diversity still seemed very close to that of biological diversity; however, the former term never took root in the same way as the latter. In 1982, the Institute of Ecology at the University of Georgia sponsored a conference called Natural Diversity in Forest Ecosystems. The conference was spurred by interest in the so-called "diversity clause" in the National Forest Management Act (NFMA) of 1976, in which, according to the preface of the proceedings, "natural diversity emerged as a requirement, criterion, and output of good forest resource management" (Cooley and Cooley 1984b: i). The papers in the proceedings explored, among other topics, "recent basic and applied research relating to natural diversity . . . [and] the legal, political, and administrative frameworks through which the concept of natural diversity is applied to land use planning and management" (Cooley and Cooley 1984b: i). It is evident, however, that many of the participants were focused on the traditional concerns of forest resource managers. For example, Orie Loucks, in the opening paper entitled "Natural Diversity as a Scientific Concept in Resource Management," suggested breaking down the "goal of sustained natural diversity" into "three manageable components: (1) diversity of landscape or cover types, (2) diversity of stand composition (the common species) and (3) diversity of uncommon species" (Loucks 1984: 1). The use of the terms *cover types* and *stand composition* reveals the author's concentration on woody plants and forest management. Other papers repeated this emphasis on trees and forestry terminology by focusing on the importance of preserving genetic resources for commercially important tree species. Indeed, an emphasis on vegetation as the guiding indicator for diversity in natural ecosystems permeated the text. For some contributors, it was apparent that the concept of natural diversity seemed to offer nothing novel beyond a new package for already established techniques of multiple-use forest management.

However, some conference participants also believed that the term *natural diversity* could have much broader implications. In fact, the legal language of the NFMA, one of the primary motivations for the conference, left open the possibility for a broad interpretation. The paragraph containing the diversity clause directed the secretary of agriculture to (among other guidelines) "provide for diversity of plant and animal communities based on the suitability and capability of the specific land area in order to meet overall multiple-use objectives, and within the multiple-use objectives of a land management plan adopted pursuant to this section, provide, where appropriate, to the degree practicable, for steps to be taken to preserve the diversity of tree species similar to that existing in the region controlled by the plan" (Peterson 1984: 22). While this passage may have been seen by some as potentially sweeping in scope, R. Max Peterson, chief of the USDA Forest Service, convincingly argued at the conference that Congress was in fact only concerned with protecting the forests of natural communities from being clear-cut and replaced with tree plantations. After providing a close reading of the congressional debates over the language of NFMA, Peterson concluded, "Congress' major intent was to discourage large scale conversions of forest types. They were not opposed to all conversions, only those that resulted in changing naturally occurring forest types over large areas. So they required diversity" (Peterson 1984: 25). Peterson emphasized that the diversity requirement, which some felt (or hoped) might pre-empt all other activities on the national forests, was only one goal in the multiple-use philosophy of the Forest Service. "Diversity is not an end in itself . . . there is no 'higher purpose' intent for diversity behind the diversity requirement of the NFMA" (Peterson 1984: 25).

Peterson's paper, however, did not take away from the apparent interest in natural diversity as a management objective. In the spirit of the 1980 CEQ chapter, a number of participants presented papers arguing that indeed diversity should be "an end in itself," and perhaps even should be considered as the guiding principle for public land managers in the United States. Even in acknowledging Peterson's interpretation of the NFMA diversity clause, some used it as a springboard for advancing a broader, complex, more preservation-oriented perspective of diversity. Hal Salwasser, Jack W. Thomas, and Fred Samson offered a slightly different analysis of the NFMA diversity mandate, in "Applying the Diversity Concept to National Forest Management." As they wrote, "Inclusion of the diversity requirement in NFMA reflected a deep-seated land ethic among the American people, a desire to live in harmony with nature . . . the inclusion of diversity as a management goal is a revolution in thought" (Salwasser, Thomas, and Samson 1984: 59–60). By the end of the paper, the authors were discussing diversity issues much more broadly than Congress's apparent intent in NFMA. "Aldo Leopold . . . proposed a land ethic in which man ensures the biotic variety and productivity of natural ecosystems while obtaining from them much needed resources. The commitment to maintain biological diversity, including viable populations of all native and desired non-native vertebrates on each national forest, is perhaps the most significant land ethic policy undertaken by a resource management agency" (Salwasser, Thomas, and Samson 1984: 66).

It is interesting to note that in extending the diversity requirement to its preservation-oriented end, Salwasser, Thomas, and Samson chose in this passage to endorse the maintenance of *biological* diversity, not *natural* diversity. Possibly the authors perceived natural diversity as having become too intertwined with traditional resource management, while biological diversity seemed more aligned with Leopold's beliefs. It was evident that Salwasser and his colleagues wanted to promote what they perceived as a new way of approaching conservation, and in so doing, they chose to overlook the likelihood that this was not the intended interpretation of those in governmental decision-making positions. Whatever Congress's interest may have been, it was evident that their choice of the term *diversity* in the NFMA—even though it was employed to protect against clear-cutting and monocultures—had stimulated a nerve in the conservation community. Salwasser and his colleagues saw *biological diversity* as a term that could be part of the "revolution in thought" that they believed was occurring.

The Wilcox Definition

At the same conference, while Salwasser and his colleagues were trying to coax their fellow conferees away from the NFMA definition of diversity toward the biological diversity notion they associated with Leopold, another contributor took a more direct definitional route. Bruce Wilcox, a conservation biologist, apparently decided that natural diversity was not a specific-enough concept and instead employed biological diversity as the alternative term of choice. His article, "Concepts in Conservation Biology: Application to the Management of Biological Diversity," began with a definition: "The growing concern over the erosion of biological diversity has prompted wide usage of 'natural diversity' and 'biological diversity' as catch-all terms to describe that quality of aspect of natural ecosystems that is being threatened or lost. The meaning of these terms is often implied but rarely stated explicitly. . . . Biological diversity can be defined generally as the diversity of life forms, the genetic diversity they contain, and the ecological functions they perform. This general definition can be expanded by further defining components of biological diversity at different levels of biological organization. . . . The different levels of biological organization are ecosystems, species, populational, and molecular. The components at each level are communities, species, populations, and genes, respectively" (Wilcox 1984: 156).

This definition, as it turns out, was the first to move beyond the two levels of genetic and ecological diversity originally identified by Norse and McManus. Wilcox was more explicit in choosing *four* "levels of biological organization" because he was trying to find a definition that would aid in guiding management decisions. As Wilcox wrote, "The above definition satisfies two important criteria for considering the practical application of the concept of biological diversity to research and management. It identifies and defines components of diversity that can serve as foci for research or management. And it recognizes interrelationships among the components at different levels of biological organization, which although complex, are sufficiently understood to allow approximate descriptions. These interrelationships and the processes that characterize them are central to understanding biological diversity in relation to its management. Further, this definition provides a conceptual framework in which practical arguments for the importance of diversity in general, for diversity at specific component levels, and for particular biological elements can be made and even tested" (Wilcox 1984: 157).

If Norse and McManus had provided a broad foundation for the concept, Wilcox was the first to give it a distinct structure. He specifically pointed to hierarchical levels of organization that served to focus both management objectives and practical arguments for conservation. While more technical in his approach, Wilcox's detailed definition constructed a framework within which people could place individual concerns about maintaining biological diversity. Norse and Mc-Manus perceived that to protect living resources, separate interests must be brought together to see the connections between genetic conservation and ecological conservation. Wilcox, by expanding the concept beyond the two related levels, was the first to make the hierarchical structure more useful.

This definition of biological diversity has rarely been cited. The article, apparently only published in these proceedings, was perhaps too obscure for the conservation community at large to notice. Wilcox would later team with Norse and others to publish the first three-tiered definition in common usage today. But the clarity, logic, and practicality of the hierarchical framework was first laid out in this paper.

Other Influential Works from the 1980s

A number of books and articles published in the first half of the 1980s provided important ingredients to the still-young concept of biological diversity, and while some only used the term in passing, each related presentation and publication fortified the groundwork supporting widespread concern for the component parts and processes of the natural world. As mentioned in chapter 1, one prominent figure in conservation at this time was Paul Ehrlich, a biologist from Stanford whose 1968 treatise on human population growth, *The Population Bomb,* had brought him fame as one of the environmental movement's most noteworthy doomsday seers. By 1981, Ehrlich and his wife, Anne, had turned their attention to the loss of species, as discussed in their influential book *Extinction: The Causes and Consequences of the Disappearance of Species.* In the preface, the Ehrlichs introduced their often-cited rivet-popping analogy: "The natural ecological systems of Earth, which supply these vital services, are analogous to the parts of an airplane that make it a suitable vehicle for human beings. But ecosystems are much more complex than wings or engines. Ecosystems, like well-made airplanes, tend to have redundant subsystems and other 'design' features that permit them to continue functioning after absorbing a certain amount of abuse. A dozen rivets, or a dozen species, might never be missed. On the other hand, a thirteenth rivet popped from a wing flap, or the extinction of a key species involved in the cycling of nitrogen, could lead to a serious accident" (Ehrlich and Ehrlich 1981: xii–xiii).

The Ehrlichs noted the primary arguments for the preservation of species—economic, ethical, and aesthetic—and dedicated (in the tradition of Norman Myers) several chapters to reviewing many of the "direct" benefits that humans would lose because of extinctions. But the focus of the book was on what the authors called the ecological or "indirect" benefits to humanity. "This argument is that other species are living components of vital ecological systems (ecosystems) which provide humanity with indispensable free services—services whose substantial disruption would lead inevitably to a collapse of civilization. By deliberately or unknowingly forcing species to extinction, *Homo sapiens* is attacking *itself;* it is certainly endangering society and possibly even threatening our own species with extermination. This is the most important of all the arguments—the one embodied in the rivet-popping analogy of our preface" (Ehrlich and Ehrlich 1981: 6). As with other texts from this time, the term biological diversity was not prominent. The Ehrlichs did discuss "genetic diversity" and its possible role in maintaining ecosystem services, and also the importance of "rainforest diversity." Still, the concerns and arguments in this book were all similar to those that would come to characterize the concept of biological diversity.

In the next year, the Ehrlichs published a short article in *Mother Earth News* that directly addressed issues of "diversity." Entitled "Saving Diversity: A Question of Habitat," the article took the broad perspective not only of preserving the whole to save the parts, but also of recognizing the web of connections in such global issues as overpopulation, acid rain, maintaining marginal lands apart from cultivation, reorganizing the economic system, closing the gap between the rich and the poor, and initiating an ethical revolution in our cultural thinking. In this same year, Paul Ehrlich published an article in *BioScience,* "Human Carrying Capacity, Extinctions, and Nature Reserves," in which he featured the maintenance of diversity as the overarching goal in saving the planet. Comparing the threat to living resources to the concern over the depletion of traditional resources such as fossil fuels, Ehrlich declared, "It is more likely that the depletion of another nonrenewable resource, the biological diversity of our planet, will be the prime factor in triggering a decline in human numbers so catastrophic that it could spell the end of industrial civilization" (Ehrlich 1982: 331). Ehrlich was never one to mince words. But in addition to such dramatic language, Ehrlich wanted to bring the issue of depleted biological diversity on par with other perceived threats to human existence. "What is required to save the remaining biotic diversity of Earth is nothing less than a dramatic paradigm shift—a transition to a 'sustainable' or 'steady-state' society. . . . Not surprisingly, the solution to this problem is intertwined with the solutions to other major human problems— economic inequity, racism, and, above all in the short term, avoidance of nuclear war. Those who claim that this is all too idealistic must carefully consider the alternatives" (Ehrlich 1982: 333). While some critics likely regarded Ehrlich's strong words as impractical hyperbole, the elevation of biological diversity into the doomsday vocabulary likely had some impact on its visibility as a newsworthy issue. The fact that the Ehrlichs chose to frame their arguments in the context of "diversity" was a telling step in the term's growth in popularity.

Norman Myers, throughout the early 1980s, continued to publish articles and books on tropical deforestation and the utilitarian benefits of living resources. His 1983 book, *A Wealth of Wild Species: Storehouse for Human Welfare,* continued the work he had documented in *The Sinking Ark,* focusing on the material value of animals, plants, and microorganisms to humans. In this book, Myers offered an expanded section on the possibilities of genetic engineering, a topic that he indicated had recently gained more widespread attention in the scientific and conservation literature: "As recently as 1980, we heard little about genetic engineering and its associated biotechnologies. A few papers in scientific journals and occasional articles in magazines—that was about it. Today we come across several reports every week. Genetic engineering has arrived" (Myers 1983: 195).

As in *The Sinking Ark,* Myers did not use the term biological diversity anywhere in *A Wealth of Wild Species,* but in the same year, in an article he authored with Edward Ayensu, "Reduction of Biological Diversity and Species Loss," he integrated the term seamlessly into his utilitarian arguments. Not unexpectedly, because of his economic emphasis, he focused on the genetic ramifications of diversity loss: "Human activities are rapidly degrading the biosphere with serious consequences for the maintenance of the planet's biological diversity in general and in particular for the preservation of its genetic resources" (Myers and Ayensu 1983: 72). Myers and Ayensu wrote specifically about the loss of wild and parental strains of important food sources and the threat to "many genetic reservoirs with startpoint materials for drugs for treatment of cancer, heart disorders, respiratory diseases, circulatory troubles, and a host of other ailments" (Myers and Ayensu 1983: 73). It was a familiar theme for Myers; his work was characterized by encyclopedic statistics on the material and economic importance of the genetic contributions of species. But most important, in the context of the umbrella aspect of the concept of biological diversity, Myers's perspective was a fitting complement to the Ehrlich's ecosystem-level perspective. The two represented distinct visions, both of which enjoyed popularity in conservation circles: one reductionistic and economically based; the other more holistic and socially informed. As both Myers and the Ehrlichs had begun to rely on the invocation of biological diversity as a primary concern, other environmentalists took notice and followed their lead, thus solidifying support for the concept that would tie their two perspectives together.

The 1985 Interagency Task Force Report: Broadening the Concept

Having captured the attention of environmental writers, natural resource managers, and national policy makers, biological diversity's next significant evolutionary step came in 1983, when the federal government established a task force of federal agencies led by the U.S. Agency for International Development. The objective of this group, established in the 1983 International Environmental Protection Act, was to write a report on current and future national efforts to conserve biological diversity both domestically and internationally. The task force report (the *U.S. Strategy on the Conservation of Biological Diversity*) was released in 1985, and like the 1981 Strategy Conference proceedings, it focused for the most part on the status of genetic resources, with specific concerns about species or ecosystems discussed in the context of protecting genetic variety. But perhaps the more significant position taken by the task force report was the integral connection it made between economic health and the conservation of living resources. The idea of sustainable development had gained many adherents in the international policy community, and biological diversity had come to represent the package of cross-disciplinary concerns that had to be considered for planning purposes. "Since biological diversity is a measure of economic potential as well as genetic wealth, the Task Force's major conclusion is that *provisions for conserving biological diversity must be incorporated into development planning*" (USAID 1985: vii; authors' emphasis). Any funding for programs concerning land-use planning, resource management, population issues, or agriculture was now perceived as a potential tool to help developing countries preserve biological diversity.

While biodiversity proponents had been talking about this perspective for several years, the breadth of the task force report's approach allowed government agencies to interpret many of their ongoing activities as already in support of biodiversity maintenance. In introducing a review of the government's efforts to date, the report claimed: "Today the United States sponsors more than 700 projects and activities that help conserve biological diversity" (USAID 1985: viii). Although the term was new, the concern for it, claimed the authors, was not. Still, there was perceived room for improvement, and the report listed sixty-seven recommendations, observing that within the established programs, "significant additional impact can be achieved through greater efforts to incorporate biological diversity concerns into existing programs," especially by enhancing "cooperation among agencies" (USAID 1985: ix).

But critics were disappointed with the lack of specificity in the report and the apparent avoidance of responsibility hinted at by the claims that present programs were already protecting biological diversity. Conservationists who had participated in the initial advisory sessions of the preparation of the report were "in general disappointed they didn't go any further" (Tangley 1985: 336). Elliott Norse, one of the members of WRI's advisory task force, called the government work "a very vague and tepid document. . . . Several good specific recommendations proposed were diluted right out" (quoted in Tangley 1985: 336). But others dismissed such statements as common environmentalist complaints. The simple fact that the government had appointed a task force that had examined the issue was seen by some conservationists as a significant step in the right direction. Peter Raven, director of the Missouri Botanical Gardens, commented, "This document is the formulation of a strategy by a government that spends several billion dollars each year on actions that have great bearing on biological diversity. . . . That alone makes it extraordinarily different from anything that has come before it" (quoted in Tangley 1985: 341).

The task force report ultimately did serve to get the focused attention of some members of Congress. Gus Yatron, chairman of the Foreign Affairs Subcommittee on Human Rights and International Organizations, declared that the preservation of biological diversity was "humankind's most fundamental environmental problem," and he held a hearing in 1985 to review the report (quoted in Wenzel 1985: 696). In his opening comments, Yatron joined the ranks of the report's critics: "I am concerned whether it really advances a cohesive strategy to address the problem and whether USAID is significantly increasing its resources toward this issue. However, the congressional mandate is clear. We need a greatly expanded program to arrest the loss of biological diversity. The magnitude of the problem is huge, but efforts to date have been sorely inadequate" (U.S. Congress 1985: 2). The hearings revealed a fundamental problem in the broad definition of biological diversity. The representatives of the environmental community, who were obviously favored at this hearing, were claiming that the task force was just a meaningless list of possible projects that the federal government could undertake but likely would not. Yatron was looking for specific improvements to the report, and in general the testimonies pointed toward traditional methods of setting aside protected areas or reserves and "building a global network of conservation units." But USAID ironically was trying to advance a much broader, more integrated view, which had been the strategy for the environmentalists in trying to gain attention for their conservation issues. As John Eriksson, deputy assistant administrator of USAID's Bureau for Science and Technology, commented, "I think it is desirable to clarify the fact that we believe that conservation of biological diversity requires a rather broad definition and that explains the breadth of the responses from our field missions when we sought their views on the strategy. . . . So our conviction . . . is that, one, biological diversity concerns cut across a wide range of sectors; two, that sustained economic development requires appropriate concern for biological diversity, and that conversely . . . the continued conservation of biological diversity is dependent upon sustained economic growth. The two go hand in hand from both sides. Three, that a large share of our development assistance efforts are already directly or indirectly supportive of the conservation of biological diversity" (U.S. Congress 1985: 67).

The problem in the eyes of conservation groups arose when concerns about the status of specific ecosystems and wildlife populations became overshadowed by the economic issues of sustainable development. Because so much of USAID's efforts already provided assistance for those economic resource concerns, the traditional targets of conservation that conservationists felt should be central to planning might have become lost in the larger scheme of resource development. As earlier noted, the foundation for the connection between sustained economic growth and biological diversity conservation had been laid by Eckholm, Myers, the World Conservation Strategy, and even the original CEQ chapter, as part of the authors' efforts to highlight the significance of living resources. The link between development and conservation was undeniable, and it had been fostered by the same community that now saw its traditional concerns being overshadowed by the broader range of economic and social problems that impacted biological diversity conservation. In a sense, concerns over the loss of biological diversity had catapulted conservation issues onto the big stage of national and international policy concerns. But the price for this attention was that they had to compete with social issues that were traditionally more important to lawmakers and their constituents.

E. O. Wilson and the National Forum on Biodiversity

While the government and the policy makers debated over the place of biodiversity conservation in relation to other global concerns, there were still numerous potential supporters outside of the government who did not yet fully comprehend the extent of the biodiversity crisis. In addition, the media had largely ignored the issue. This fact was not lost upon many of the proponents of diversity conservation, in particular members of the scientific community who perceived that the losses of biological diversity deserved more serious attention from scientists and citizens alike. In 1985, the same year that the task force report was published and the congressional hearings held, E. O. Wilson published his seminal article "The Biological Diversity Crisis: A Challenge to Science" (see chapter 1), and the National Academy of Sciences

(NAS) began to plan the National Forum on Biodiversity. One of the lead organizers of the event was Walter G. Rosen, a senior program officer from the National Research Council serving on its Board of Basic Biology. Rosen recalls in an interview that the topic "that kept rising to the surface" at board meetings "was the loss of biological diversity, the increasing frequency of extinctions" (Takacs 1996: 36). Rosen took the issue to the NAS and proposed a national forum on biological diversity loss. The Smithsonian was later brought in as co-sponsor, and the event gained momentum in both the scientific community and the media.

In fact, even though the convention was planned under the auspices of the notably unbiased NAS, it seemed equally as focused on "spreading the word" about the crisis of biodiversity as on the scientific aspects of its decline. Rosen called the forum an "exercise in consciousness raising. . . . Although the biological diversity problem had penetrated the consciousness of scientists, there was really not much public awareness of it" (quoted in Tangley 1986: 708). An article in *BioScience,* "Biological Diversity Goes Public," focused on the fact that the conference was making a huge splash with the press: "The forum also attracted more than 40 journalists and generated major stories in *Time* magazine, *The Washington Post, The New York Times, The Boston Globe,* and other daily newspapers around the country" (Tangley 1986: 708). David Takacs, in his book *The Idea of Biodiversity* (1996), notes that such publicity was different from that to which the traditionally conservative and objective NAS was accustomed. Partially, Takacs believes, the wide publicity resulted from the Smithsonian's participation and its more media-friendly methods of planning and presenting conferences. But Takacs also points out that this was an event at which scientists and others came together for the express purpose of raising awareness of an issue about which they felt strongly. The central concept had even acquired the media-ready name *biodiversity,* which Rosen apparently had coined partially for conference publicity purposes. When Takacs asked E. O. Wilson about the term, Wilson replied that he had first opposed it as " 'too glitzy' but . . . now admits that Rosen 'turned out to be completely right.' " Takacs continues, "The glitziness of the word contributed to the speed of its spread, and Wilson told me that he is pleased that biodiversity is now 'thoroughly ensconced' " (in Takacs 1996: 37).

But perhaps most important, by being so successful as a consciousness-raising event, the forum brought the loss of biological diversity out of the narrow purview of environmental organizations and government agencies and made it an issue of national concern. Interest in biological diversity had spread beyond its traditional home in conservation and the sciences. As Wilson remarked to Takacs, "The forum came to be not just about the biology of the origination of diversity and extinction, but also all of the other concerns, through ecology, population biology, and in the most novel development, economics, sociology, and even the humanities. So in one stroke, the biology and the focus of biodiversity was recognized as a concern of a large array of disciplines" (quoted in Takacs 1996: 39). The combined publicity and wide appeal of the concept allowed it to spread to all corners of the conservation community. Wilson would later edit the papers presented at the forum in a seminal collection, *Biodiversity* (1988). The book marked a watershed event. As Wilson wrote, "It . . . documents a new alliance between scientific, governmental, and commercial forces—one that can be expected to reshape the international conservation movement for decades to come" (Wilson 1988: vi).

Wilson also noted that he believed the increased attention to biological diversity issues could be attributed to "two more or less independent developments. The first was the accumulation of enough data on deforestation, species extinction, and tropical biology to bring global problems into sharper focus and warrant broader public exposure. . . . The second development was the growing awareness of the close linkage between conservation of biodiversity and economic development" (Wilson 1988: v–vi). Wilson's first observation suggests that the information collected on degraded habitats and declining species numbers had finally reached a critical mass, not only to mobilize the scientific community to organize in response to the crisis, but also to spill over into the public arena with force. Wilson's "second development," the "close linkage between conservation of biodiversity and economic development," was the discussion that had dominated the government reports up to the time of the forum, and it was this economic connection that gave the call to protect biodiversity the added weight it needed to be included in discussions across disciplines. Indeed, Norse and McManus's vision of bringing together different interests under the aegis of one unifying goal was practically realized. While one could see weaknesses in the all-encompassing nature of the concept of biological diversity, particularly in applications of it to analyses of how the government might implement conservation plans, this very breadth revealed itself as valuable in the opportunities that arose for communicating the pervasiveness and severity of the problem. As Wilson's comments imply, there was something in this concept for everyone, and the data and ideas on biological diversity had matured enough so that people could latch on to the issues and relate them to their own concerns. The media recognized this characteristic of *biodiversity* (as did the forum's promoters), and the concept came to be firmly established in the vocabulary of those promoting any aspect of the natural environment. The breadth of the concept in the public arena had served it well.

Conclusion

In the same year as the forum, Norse and his colleagues published the first three-tiered definition of biodiversity in *Conserving Biological Diversity in Our National Forests.* This definition was in turn solidified in the publication *Technologies to Maintain Biological Diversity,* released by the Office of Technology Assessment in 1987. But well before this, people were considering the protection of biological diversity on the multiple levels of species, genes, and ecosystems, both in language and in practice. Norse himself recounts that even as early as 1981, immediately following the publication of the first definition, he was making presentations that went beyond the original two levels to include ecosystem diversity in the concept (Norse 1999). Bruce Wilcox's 1982 four-tiered definition, while not widely recognized, showed that some conservationists were following the examples of those like Eckholm and the authors of the World Conservation Strategy in approaching the protection of the natural world from a multiple-level perspective. The three levels that became integrated into the standard definition represented the most commonly recognized components of nature. Norse's three-level definition in 1986 was more an encapsulation of the discussions that had occurred over the previous decade than a revolutionary step in the development of the concept.

While this definition of biological diversity was firmly established by the late 1980s, the debate over the efficacy of biodiversity conservation efforts in the context of sustainable development continued. It could easily be argued that biodiversity's broad definition—especially its connection to economic value—invited those more interested in development to co-opt the term and use it to advance exploitive policies over protectionist goals. Others declare that conservation issues had never enjoyed equal consideration with development issues, and that the joining of these opposites was the most important step of all. Whichever perspective one chooses, it is difficult to deny that biological diversity represented a coming together of interests that had long been searching for a common objective. By uniting the conservation world in this way, the concept gained enough momentum to increase awareness of the degradation of the natural world among constituencies that had not previously shown much interest in environmental concerns. Most important, it answered a need that the conservation community had been trying to articulate for at least a decade. As Elliott Norse commented in an interview, "The concept had real value. . . . It looked at conservation differently. It had the potential to bring people together who had not been together before. . . . It was an idea whose time had come" (Norse 1999).

Coda

Elliott Norse, writing in 1996, used the following analogy to characterize how intellectual and cultural movements grow from and build upon multiple sources: "A movement in human society is like a river system with many beginnings, in which tiny drops of water form and coalesce within tiny catchments as rivulets that join within progressively larger catchments as small streams, larger creeks, and rivers that ultimately flow to the sea" (Norse 1996: 5). Norse evocatively lays out the task that faces historians who are interested in tracing the evolution of a particular idea. The process of creating a map that represents the many influences that have contributed to the development of a concept can be a daunting challenge. Inevitably, some will feel that certain details have been overlooked, or perhaps that some streams of thought have been unduly emphasized over others. My goal in writing this history of the concept of biological diversity has been to try to present a broad understanding of the cultural sources of our concern for and fascination with the variety of life on earth. As such, I have no doubt omitted numerous noteworthy publications, personalities, and events that have played some contributing role in the history of the concept. But it is my hope that in my selection of sources I have been able to provide an accurate outline of the major tributaries that flowed together to form the concern for living diversity in its many aspects. Even more generally, I believe that it is instructive for those in the conservation world to occasionally step back and examine the assumptions and values that underlie the causes that define their work. By seeing the conservation of biological diversity as part of a long continuum of environmental concerns, we are reminded that the drive to protect the natural world has a long tradition in our society, and we can perhaps draw upon the knowledge of past efforts to inspire those of the present and future.

The two organizational frameworks that were used—the definitional hierarchical levels and a vocabulary of environmental values—proved extremely useful for building the historical narrative. The respective categories of species, genes, and ecosystems were "natural" divisions within the literature, each having a clear set of foundational sources from which to proceed in researching historical concerns for the natural world. This divisional quality supported the idea that the definition of biological diversity evolved to include these three levels because they were the most prominent and distinctive interests in the conservation and scientific worlds. By drawing upon the various constituencies that traditionally were concerned with protecting species, genes, and ecosystems, promoters of the biodiversity concept ensured a strong and lasting support base.

The values vocabulary provided a way to illustrate significant developments in the separate conservation histories and to show how a broadening awareness of the benefits that humans reap from the natural world helped to formulate support for the development of the concept of biological diversity. Each history is characterized by a widening recognition of values as expressed in the literature, generally focusing on utilitarian and scientific values in earlier years, then expanding to include all other values in various forms through the decades of the twentieth century. In addition, as more values gained currency, the links between the different environmental concerns became more apparent. For example, as the ecological benefits of intact natural systems became more valued, the role of species diversity in maintaining ecosystem stability became a key point of interest; as the potential utilitarian value of genetic diversity became more obvious, the drive to protect ecosystems and species that harbored valuable genes became an imperative; and as the scientific interest in preserving a representative sample of the world's ecosystems became more widespread, it necessarily included the need to protect both genetic variety and species diversity, such that ecosystems could be studied in their original condition for many years to come. These connections were based in the growing awareness of the interdependency of the natural world, and their recognition was likely enhanced by the fact that people were perceiving certain values in the human relationship with living nature that had not previously been broadly acknowledged. In this way, the convergence of the concern for the three levels and the full range of environmental values laid the foundation for an overarching concept. With the general growth of interest in diversity over the 1960s and 1970s, the conservation community was well prepared to consider a diversity-based objective that would serve to encompass the previously disparate concerns for protecting the natural world.

While the history that led to the popularity of the concept provided the central focus of this book, it was equally interesting to examine how the definition of biological diversity was shaped through the term's subsequent usage after the first published definition in 1980. Norse and McManus's original association of biological diversity with its utilitarian value was a distinctive characteristic in which many were interested, particularly those connected to the conferences and reports sponsored by the U.S. federal government in the early 1980s. The early connection of the conservation of "living resources" to sustainable development—as expressed in the *World Conservation Strategy,* for example—also impacted many debates and discussions within conservation since 1980.

Members of the conservation and scientific communities have worked hard to promote the inclusive character of biological diversity, emphasizing in particular the ecological and scientific values of the natural world. Paul Ehrlich and others drew support from the diversity-stability debate, focusing on biodiversity's contributions to ecosystem services, while E. O. Wilson and scientists who participated in the National Forum defended conservation practices because of the high scientific value of the diversity of life. The other values associated with protecting biodiversity also found various forms of expression in the 1980s. For example, the growing popularity of ecotourism highlighted the recreational and aesthetic value of the natural world. The moral imperative associated with the founding of the Society for Conservation Biology in 1986 acknowledged the ethical value of protecting the variety of life on earth. The recognition of "national sovereignty" over biological resources in the 1992 Convention on Biological Diversity partially recognized the cultural value of a healthy biota to its home nation. In short, the umbrella aspect originally envisioned in Norse and McManus's CEQ chapter was realized over a number of years. By allowing the framework of biological diversity conservation to be so broad, the authors and subsequent promoters provided enough variety for numerous interests to find a niche within the concept. When the term began to catch on, there was ample room "on the bandwagon" for potential supporters to climb aboard. The concept of biological diversity thus gained adherents and grew in popularity over the 1980s and 1990s, developing into a pervasive theme.

But the success of the biological diversity paradigm has not come without criticism. There are many who believe that biodiversity's inclusiveness has been its greatest weakness, as opposed to being the source of its strength. Because so many interests perceive a stake in conserving biological diversity, some believe that the concept has become unnecessarily diluted and that the central cause of protecting the natural world has been overshadowed. In particular, critics have expressed concern over the emphasis of biodiversity's utilitarian value over other benefits, such as those related to scientific or ethical values. The conflict has roots dating back to the original 1981 U.S. Strategy Conference, when discussions of the use value of genetic diversity dominated presentations. More recently, the utilitarian value of biodiversity has shifted its focus from genetic variety to the importance of ecosystem services. While this may seem like a welcome change in perspective, some argue that the new emphasis has more to do with estimating how far an ecosystem can be pushed by development interests while still providing certain services to humans, and less to do with maintaining healthy, intact populations and ecosystems. This trend in conservation has been particularly evident in developing tropical countries, where biodiversity is richest and the people often very poor. Randall Kramer and Carel van Schaik, co-editors (along with Julie Johnson) of a book entitled *Last Stand: Protected Areas and the Defense of Tropical Biodiversity* (1997), note that the new management strategies are largely influenced by the rising interest in combining biodiversity conservation with the goals of sustainable development: "During the past few years, attempts to link rain forest protection with sustainable development have led to a noticeable expansion of the meaning of the phrase 'biodiversity conservation.' In this increasingly popular view, biodiversity has come to represent ecological services and products such as clean air and water. This definition has led to a shift away from species protection and towards sustainable use" (Kramer and van Schaik 1997: 4). Thus, the dangers of such a broad concept are revealed in constituents and activities with which the term has become associated. Interests that wish to protect biodiversity mainly for human utility are not the traditional preservation interests that have fought for biodiversity protection in spite of nature's use values. However, the sustainable use supporters are also connected to the history of the concept, and their interests should at least be considered alongside those of the more preservation-oriented constituents.

In short, both groups claim to be working toward the same objective: the conservation of biological diversity. But to conservationists like Kramer and van Schaik, this is not a proper use of the term and concept. As they observe, "This dilution of the biodiversity concept has been very effective in increasing the appeal of conservation efforts to a use-oriented audience. . . . This utilitarian perspective leads to a conservation strategy in which there is little or no need for strictly protected areas. In contrast, the species [protectionist] perspective, which emphasizes the ethical and scientific values of biological species regardless of their utility for humans, leads to an emphasis on protection of examples of natural ecosystems, little modified by human action" (Kramer and van Schaik 1997: 4).

This conflict in the conservation community has led to what Peter Wilshusen and his colleagues (2002) call a resurgence of a "protection paradigm." They note that several recent publications, including *Last Stand,* have voiced strong support for restricted park systems, particularly in tropical areas where resource use by local peoples has been allowed in an effort to combine biodiversity conservation with sustainable development practices. Wilshusen and his colleagues offer a critique of this protectionist position and present numerous counterarguments supporting the idea that some development and use of the resource base must be allowed. As the authors conclude, "In general should we treat rural people as potential allies or as potential enemies? In the end, we have two broad choices. We can promote a policy shift toward authoritarian protectionism that would most likely alienate key allies at local, regional and national levels and thus precipitate resistance and conflict. Alternatively, we can build on past experience and constructively negotiate ecologically sound, politically feasible and socially just programs in specific contexts that can be legitimately en-

forced based on strong agreements with all affected parties" (Wilshusen et al. 2002: 35–36). However, as with many supporters of sustainable development, these authors are sympathetic to those who see
biodiversity protection as being severely compromised. In addition to
the frustration of the conflicting parties, it is confusing to those within
and outside of the discussion when both sides claim to be concerned
with "biodiversity conservation."

Ironically, because of the breadth of the concept, and because of its
close historical association with sustainable development, neither side
can claim that the other is misusing the term. This observation raises
some intriguing questions. For example, for all its pervasive use and
popularity, is the biodiversity concept a useful framework for actual
conservation practices? Some critics maintain that biological diversity
has simply become a synonym for "nature" or "the environment," and
as such can be used or misused by anybody whose interests somehow
involve components of the natural world. Does protecting biodiversity
imply certain management objectives and the predominance of certain
values over others? Protectionists like Kramer and van Schaik would
likely favor ecological, scientific, and ethical values over utilitarian values. Sustainable use supporters like Wilshusen and his colleagues believe that some kind of compromise is essential if a lasting balance
between the needs of human populations and the conservation of
biodiversity is the objective. Is either side more faithful to the original
intent of the concept?

The answers to these questions lie in the fact that the concept of biological diversity was never intended to give direction to solving specific conservation problems. It was introduced as an umbrella concept
and was received and promoted by the conservation, scientific, and
policy-making communities as an overarching multilevel perspective
of the natural world and the different ways we value it. As such, biological diversity provides only a general guide to conservation efforts
—specifically, that in making decisions that impact living resources,
one must be cognizant of the effects on all levels and all values. The
concept provides no direction as to which values should be favored
over others. Rather, it requires that all values be recognized—the inclusiveness that is the main reason why biodiversity conservation has
become so closely associated with sustainable development.

The major contribution of the concept of biological diversity has been its ability to bring together previously separate concerns and unite them under one objective. In the 1970s, the conservation community was seeking a unifying theme, and diversity in the natural world was increasingly characterized as a normative good. When Norse and McManus introduced the concept in 1980, they were able to build upon earlier uses of the term and the general need for a broad perspective that could include the multiple interests and values associated with conserving nature. Because of its ability to gain support from a wide variety of constituents, biological diversity advanced conservation causes further than ever before. Concern for living resources was considered on the same level as concern for human rights, political strategic interests, and global economic considerations.

As for the application of conservation on the ground, the biodiversity concept gives little specific guidance. Because it emphasizes both protection and use, managers and decision makers are left to consider the merits of each individual issue. This quality has left many frustrated with the questionable usefulness of the biodiversity concept. But its implicit internal conflict encapsulates the root dilemma of the conservation movement. From the competing perspectives of conservationist Gifford Pinchot and preservationist John Muir, to the A/B cleavage cited by Aldo Leopold, to the debate over sustainable development, conservationists have had to contend with finding the appropriate balance between the human use and consumption of environmental goods and the nonuse values inherent in protecting the natural world from adverse impacts. The role of biological diversity in this history is one of mediator, promoter, and catalyst. The concept has brought interests together, raised the general awareness of several important interrelated issues, and inspired numerous conferences, publications, and organizational programs that worked to solve the environmental problems facing both human and nonhuman life. Biological diversity was never envisioned as providing objectives. Its contribution has been to encourage people to think inclusively when making conservation decisions.

There are those who claim that as a unifying concept, biological diversity has outlived its usefulness. As the sustainable use controversy illustrates, the biodiversity cause could be seen as creating more divisions than partnerships in the conservation world. But there are certain characteristics that suggest that the biological diversity concept may have some significant lasting power. Foremost, as detailed in this book, the concept has deep-seated roots that tie it to various conservation traditions. Its coverage of the full range of environmental values ensures a broad support base. It carries a scientific authority that impresses people both inside and outside of the scientific community. One potential drawback is that it seems only a small percentage of people understand the meaning of the term. Some claim that this lack of recognition with the public proves the term is too esoteric to be effective in the future. But it is equally possible that the concept is simply taking time to filter through the general population, and because of its broad support base it seems unlikely that the term will disappear from environmentalism's vocabulary. Certainly, another concept will someday become the new cause for conservation. Our society constantly needs to renew ideas, and the discourse we choose is the symbolic carrier of those ideas. But the paradigm of thinking about the living world as biological diversity is firmly embedded in our collective environmental consciousness, and it will likely be some time before a new paradigm pushes it aside.

References

Adams, Charles C. 1925. "The Conservation of Predatory Mammals." *Journal of Mammalogy* 6: 83–96.

Allee, W. C., Alfred E. Emerson, Orlando Park, Thomas Park, and Karl P. Schmidt. 1949. *Principles of Animal Ecology*. Philadelphia: W. B. Saunders.

Allen, Durward. 1954. *Our Wildlife Legacy*. New York: Funk and Wagnalls.

———. 1973. "Needed: A New North American Wildlife Policy." *National Parks and Conservation Magazine* 47 (August): 10–13.

Allen, John Morgan. 1963. *The Nature of Biological Diversity*. New York: McGraw-Hill.

Allen, Robert. 1960. "Do We Want to Save the Whooping Crane?" *Audubon Magazine* 62 (May–June): 122–125+.

Ashe, W. W. 1922. "Reserved Areas of Principal Forest Types as a Guide in Developing an American Silviculture." *Journal of Forestry* 20 (October): 276–283.

Baker, John H. 1954. "Saving Man's Wildlife Heritage." *National Geographic* 106, no. 5: 581–620.

Baldwin, Henry I. 1941. "An Inventory of the Natural Vegetation Types and the Need for Their Preservation." *Science* 93, no. 2404: 81–82.

Barker, Will. 1956. "Our Vanishing Species." *American Forests* (March): 14–16.

Bean, Michael J., and Melanie J. Rowland. 1997. *The Evolution of National Wildlife Law*. 3rd ed. Westport, CT: Praeger.

Behlen, Dorothy. 1981. "Taking Root." *Nature Conservancy News* 31, no. 4: 7–11.

Biffen, R. H. 1906. "Mendel's Laws of Inheritance in Wheat Breeding." *Journal of Agricultural Sciences* 1: 4–48.

Black, John D. 1954. *Biological Conservation, with Particular Emphasis on Wildlife*. New York: Blakiston.

Blair, W. Frank. 1977. *Big Biology: The US/IBP*. Stroudsburg, PA: Dowden, Hutchinson and Ross.

Boerma, A. H. 1970. Foreword to *Genetic Resources in Plants: Their Exploration and Conservation,* edited by Otto H. Frankel and Erna Bennett. Philadelphia: F. A. Davis.

Bormann, F. Herbert, and Stephen Kellert. 1991. *Ecology, Economics, Ethics: The Broken Circle.* New Haven: Yale University Press.

Brown, William L. 1982. "Biological Diversity: Basic for Agricultural Success." In *Proceedings of the U.S. Strategy Conference on Biological Diversity, November 16–18, 1981,* U.S. Agency for International Development. Washington, DC: Agency for International Development.

———. 1984. "Conservation of Gene Resources in the United States." In *Plant Genetic Resources: A Conservation Imperative,* edited by Christopher W. Yeatman, David Kafton, and Garrison Wilkes. Boulder, CO: Westview.

Buchheister, Carl W. 1961. "The Wilderness Bill and Endangered Wildlife." *Audubon Magazine* 63 (May–June): 153+.

Buckman, Robert E., and Richard L. Quintus. 1972. *Natural Areas of the Society of American Foresters.* Washington, DC: Society of American Foresters.

Carson, Rachel. 1962. *Silent Spring.* Boston: Houghton Mifflin.

Cart, Theodore Whaley. 1971. "The Struggle for Wildlife Protection in the United States, 1870–1900: Attitudes and Events Leading to the Lacey Act." Ph.D. diss., University of North Carolina, Chapel Hill.

Caughley, G. 1970. "Eruption of Ungulate Populations, with Emphasis on Himalayan Thar in New Zealand." *Ecology* 51, no. 1: 53–72.

Clements, Frederic E., and Victor E. Shelford. 1939. *Bio-ecology.* New York: John Wiley and Sons.

Cooley, James L., and June H. Cooley, eds. 1984a. *Natural Diversity in Forest Ecosystems.* Athens: Institute of Ecology, University of Georgia.

———. 1984b. Preface to *Natural Diversity in Forest Ecosystems,* edited by James L. Cooley and June H. Cooley. Athens: Institute of Ecology, University of Georgia.

Coolidge, Harold. 1963. Introduction to *Scientific Use of Natural Areas Symposium,* edited by Julia Field and Henry Field. Washington, DC: International Congress of Zoology.

Cowan, Ian McTaggert. 1966a. "Introductory Statement and Summary Remarks." In *Future Environments of North America,* edited by F. Fraser Darling and John P. Milton. Garden City, NY: Natural History Press.

———. 1966b. "Management, Response, and Variety." In *Future Environments of North America,* edited by F. Fraser Darling and John P. Milton. Garden City, NY: Natural History Press.

Croker, Robert A. 1991. *Pioneer Ecologist: The Life and Work of Victor Ernest Shelford, 1877–1968.* Washington, DC: Smithsonian Institution Press.

Crowe, Philip K. 1970. *World Wildlife: The Last Stand.* New York: Charles Scribner's Sons.

Darling, F. Fraser. 1966. Introduction to *Future Environments of North America,* edited by F. Fraser Darling and John P. Milton. Garden City, NY: Natural History Press.

Darling, F. Fraser, and John P. Milton, eds. 1966. *Future Environments of North America*. Garden City, NY: Natural History Press.

Darnell, Reznear M. 1976. "Natural Areas Preservation: The US/IBP Conservation of Ecosystems Program." *BioScience* 26, no. 2: 105–108.

Dasmann, Raymond F. 1968. *Environmental Conservation*. 2nd ed. New York: John Wiley and Sons.

Devoe, Alan. "On Salvaging Nature." 1944. *American Mercury* 59, no. 249: 366–369.

Diamond, Jared. 1976. "Island Biogeography and Conservation: Strategy and Limitations." *Science* 193, no. 4257: 1027–1029.

Dice, Lee R. 1925. "The Scientific Value of Predatory Mammals." *Journal of Mammalogy* 6, no. 1: 25–27.

———. 1943. *The Biotic Provinces of North America*. Ann Arbor: University of Michigan Press.

———. 1952. *Natural Communities*. Ann Arbor: University of Michigan Press.

DiSilvestro, Roger L. 1993. *Reclaiming the Last Wild Places: A New Agenda for Biodiversity*. New York: Wiley.

Doyle, Jack. 1986. *Altered Harvest: Agriculture, Genetics, and the Fate of the World's Food Supply*. New York: Penguin.

Dunlap, Thomas R. 1988. *Saving America's Wildlife*. Princeton, NJ: Princeton University Press.

Dunn, L. C. 1965. *A Short History of Genetics: The Development of Some of the Main Lines of Thought, 1864–1939*. New York: McGraw-Hill.

Eckholm, Erik. 1978. *Disappearing Species: The Social Challenge*. Worldwatch Paper 22. Washington, DC: Worldwatch Institute.

Egler, Frank E. 1941. "Establishment of a Natural Area on the Huntington Wildlife Forest." *Science* 94, no. 2427: 16–17.

Ehrenfeld, David. 1970. *Biological Conservation*. New York: Holt, Rinehart, and Winston.

Ehrlich, Paul. 1968. *The Population Bomb*. New York: Ballantine.

———. 1982. "Human Carrying Capacity, Extinctions, and Nature Reserves." *BioScience* 32, no. 5: 331–333.

Ehrlich, Paul, and Anne Ehrlich. 1981. *Extinction: The Causes and Consequences of the Disappearance of Species*. New York: Random House.

———. 1982. "Saving Diversity: A Question of Habitat." *Mother Earth News* (September–October): 150–151.

Eisner, Thomas. 1983. "Chemicals, Genes, and the Loss of Species." *Nature Conservancy News* 33, no. 6: 23–24.

Eldredge, Niles. 1998. *Life in the Balance: Humanity and the Biodiversity Crisis*. Princeton, NJ: Princeton University Press.

Elliott, Hugh, ed. 1972. *Second World Conference on National Parks*. Morges, Switzerland: International Union for Conservation of Nature and Natural Resources.

Elton, Charles S. 1927. *Animal Ecology*. New York: Macmillan.

————. 1958. *The Ecology of Invasions by Animals and Plants.* New York: John Wiley and Sons.

Erwin, Terry L. 1983. "Tropical Forest Canopies: The Last Biotic Frontier." *Bulletin of the Entomological Society of America* 29, no. 1: 14–19.

Farnsworth, Edward G., and Frank B. Golley. 1973. *Fragile Ecosystems: Evaluation of Research and Applications in the Neotropics.* New York: Springer-Verlag.

Federal Committee on Ecological Reserves (FCER). 1977. *A Directory of Research Natural Areas on Federal Lands of the United States of America.* Washington, DC: United States Forest Service.

Fisher, James, Noel Simon, and Jack Vincent. 1969. *Wildlife in Danger.* New York: Viking.

Food and Agriculture Organization (FAO). 1985. *FAO: The First Forty Years.* Rome: FAO.

Fowler, Cary, and Pat Mooney. 1990. *Shattering: Food, Politics, and the Loss of Genetic Diversity.* Tucson: University of Arizona Press.

Fox, Stephen. 1981. *John Muir and His Legacy: The American Conservation Movement.* Boston: Little, Brown.

Frankel, Otto H. 1970a. "Genetic Conservation in Perspective." In *Genetic Resources in Plants: Their Exploration and Conservation,* edited by Otto H. Frankel and Erna Bennett. Philadelphia: F. A. Davis.

————. 1970b. "Genetic Conservation of Plants Useful to Man." *Biological Conservation* 2, no. 3: 162–169.

————. 1983. Foreword to *Genetics and Conservation: A Reference for Managing Wild Animal and Plant Populations,* edited by Christine M. Schonewald-Cox, Steven M. Chambers, Bruce MacBryde, and W. Lawrence Thomas. Menlo Park, CA: Benjamin/Cummings.

Frankel, Otto H., and Erna Bennett. 1970a. "Genetic Resources." In *Genetic Resources in Plants: Their Exploration and Conservation,* edited by Otto H. Frankel and Erna Bennett. Philadelphia: F. A. Davis.

————, eds. 1970b. *Genetic Resources in Plants: Their Exploration and Conservation.* Philadelphia: F. A. Davis.

Frankel, Otto H., and J. G. Hawkes. 1975. "Genetic Resources: The Past Ten Years and the Next." In *Crop Genetic Resources for Today and Tomorrow,* edited by Otto H. Frankel and J. G. Hawkes. Cambridge: Cambridge University Press.

Frankel, Otto H., and Michael E. Soulé. 1981. *Conservation and Evolution.* Cambridge: Cambridge University Press.

Gabrielson, Ira N. 1942. *Wildlife Conservation.* New York: Macmillan.

Gaston, Kevin J. 1996. *Biodiversity: A Biology of Numbers and Difference.* Cambridge, MA: Blackwell Science.

Goddard, Donald, ed. 1995. *Saving Wildlife: A Century of Conservation.* New York: Harry N. Abrams, in association with the Wildlife Conservation Society.

Goldman, E. A. 1925. "The Predatory Mammal Problem and the Balance of Nature." *Journal of Mammalogy* 6, no. 1: 28–33.

Golley, Frank B. 1993. *A History of the Ecosystem Concept in Ecology.* New Haven: Yale University Press.

Gomez-Pompa, A., C. Vazquez-Yanes, and S. Guevara. 1972. "The Tropical Rain Forest: A Nonrenewable Resource." *Science* 177, no. 4051: 762–765.

Goodman, Daniel. 1975. "The Theory of Diversity: Stability Relationships in Ecology." *Quarterly Review of Biology* 50, no. 3: 237–266.

Graham, Edward H. 1947. "Wildlife Is Part of Our Heritage." *Audubon Magazine* 49 (March): 96–105.

Grinnell, George Bird, and Charles Sheldon, eds. 1925. *Hunting and Conservation.* New Haven: Yale University Press.

Groombridge, Brian, ed. 1992. *Global Biodiversity: Status of the Earth's Living Resources; A Report.* New York: Chapman and Hall.

Guggisberg, C. A. W. 1970. *Man and Wildlife.* New York: Arco.

Hagen, Joel B. 1992. *An Entangled Bank: The Origins of Ecosystem Ecology.* New Brunswick, NJ: Rutgers University Press.

Haig, I. T. 1941. "The Preservation of Natural Areas Exemplifying Vegetation Types." *Science* 94, no. 2433: 163.

Hanson, Herbert C. 1939. "Check-Areas as Controls in Land Use." *Scientific Monthly* 48, no. 2: 130–146.

Hardin, Garret. 1969. "Not Peace, but Ecology." In *Diversity and Stability in Ecological Systems,* Brookhaven Symposium of Biology 22, edited by G. M. Woodwell and H. H. Smith. Upton, NY: Brookhaven National Laboratory.

Harlan, H. V., and M. L. Martini. 1936. *U.S. Department of Agriculture Yearbook of Agriculture, 1936.* Washington, DC: Government Printing Office.

Harlan, Jack R. 1961. "Geographic Origin of Plants Useful to Agriculture." In *Germ Plasm Resources,* edited by Ralph E. Hodgson. Washington, DC: American Association for the Advancement of Science.

———. 1975. "Our Vanishing Genetic Resources." *Science* 188, no. 4188: 618–621.

Hawksworth, D. L. 1995. *Biodiversity: Measurement and Estimation.* London: Chapman and Hall.

Hays, Samuel P. 1979. *Conservation and the Gospel of Efficiency: The Progressive Conservation Movement, 1890–1920.* 2nd ed. New York: Atheneum.

———. 1987. *Beauty, Health, and Permanence: Environmental Politics in the United States, 1955–1985.* New York: Cambridge University Press.

Holden, Constance. 1974. "Scientists Talk of the Need for Conservation and an Ethic of Biotic Diversity to Slow Species Extinction." *Science* 184, no. 4137: 646–647.

Holdgate, Martin. 1999. *The Green Web: A Union for World Conservation.* London: Earthscan.

Hornaday, William T. 1913a. *Our Vanishing Wild Life: Its Extermination and Preservation.* New York: Charles Scribner's Sons.

———. 1913b. *Thirty Years War for Wildlife.* Stamford, CT: Permanent Wildlife Protection Fund.

Hough, A. F. 1941. "Natural Area Established in Northwestern Pennsylvania Virgin Forest." *Ecology* 22, no. 1: 85–86.

Hunter, Malcolm. 1996. *Fundamentals of Conservation Biology.* Cambridge, MA: Blackwell Science.

Hutchinson, G. Evelyn. 1940. Review of *Bio-ecology,* by Frederic E. Clements and Victor E. Shelford. *Ecology* 21, no. 2: 267-268.

———. 1959. "Homage to Santa Rosalia; or, Why Are There So Many Kinds of Animals?" *American Naturalist* 93 (May–June): 145–159.

Hyland, Howard L. 1984. "History of Plant Introduction in the United States." In *Plant Genetic Resources: A Conservation Imperative,* edited by Christopher W. Yeatman, David Kafton, and Garrison Wilkes. Boulder, CO: Westview.

International Union for the Conservation of Nature and Natural Resources (IUCN). 1966. *Red Data Book.* Morges, Switzerland: IUCN.

———. 1980. *World Conservation Strategy: Living Resource Conservation for Sustainable Development.* Gland, Switzerland: IUCN-UNEP-WWF.

———. 1982. Commission on National Parks and Protected Areas. *The World's Greatest Natural Areas: An Indicative Inventory of Natural Sites of World Heritage Quality.* Gland, Switzerland: IUCN.

Jenkins, Robert E. 1972a. "What *Is* a Preserve?" *Nature Conservancy News* 22, no. 1: 17–18.

———. 1972b. "A Natural Areas Inventory." *Nature Conservancy News* 22, no. 3: 16–18.

———. 1973. "Why Save Land?" *Nature Conservancy News* 23, no. 1: 16–17.

———. 1974. "Land Use and the Conservancy." *Nature Conservancy News* 24, no. 1: 20–21.

———. 1975. "Heritage Programs: Inventory Progress Report." *Nature Conservancy News* 25, no. 2: 26–27.

———. 1978. "Heritage Classification: The Elements of Ecological Diversity." *Nature Conservancy News* 28, no. 1: 24–30.

Jenkins, Robert E., and Helmut P. Moyseenko. 1976. "Progress in Developing a National Database on Ecological Preservation." *Nature Conservancy News* 26, no. 1: 25–26.

Kellert, Stephen R. 1979. *Public Attitudes toward Critical Wildlife and Natural Habitat Issues.* Washington, DC: Government Printing Office.

———. 1980a. *Activities of the American Public Relating to Animals.* Washington, DC: Government Printing Office.

———. 1980b. "Contemporary Values of Wildlife in American Society." In *Wildlife Values,* edited by William W. Shaw and Ervin H. Zube. Tucson, AZ: Center for Assessment of Noncommodity Natural Resource Values.

———. 1996. *The Value of Life: Biological Diversity and Human Society.* Washington, DC: Island Press/Shearwater.

———. 1997. *Kinship to Mastery: Biophilia in Human Evolution and Development.* Washington, DC: Island Press.

Kellert, Stephen R., and Edward O. Wilson. 1993. *The Biophilia Hypothesis.* Washington, DC: Island Press.

Kellert, Stephen R., and Miriam Westervelt. 1981. *Trends in Animal Use and Perception in Twentieth-Century America.* Washington, DC: Government Printing Office.

Keystone Center. 1991. *Biological Diversity on Federal Lands: A Report of a Keystone Policy Dialogue.* Washington, DC: Keystone Center.

King, Ralph. 1947. "The Future of Wildlife in Forest and Land Use." In *Transactions of the Twelfth North American Wildlife Conference,* edited by Ethel M. Quee. Washington, DC: Wildlife Management Institute.

Kloppenburg, Jack Ralph. 1988. *First the Seed: The Political Economy of Plant Biotechnology, 1492–2000.* New York: Cambridge University Press.

Kormondy, Edward. 1965. *Readings in Ecology.* Englewood Cliffs, NJ: Prentice-Hall.

Kramer, Randall A., and Carel van Schaik. 1997. "Preservation Paradigms and Tropical Rainforests." In *Last Stand: Protected Areas and the Defense of Tropical Biodiversity,* edited by Randall A. Kramer, Carel van Schaik, and Julie Johnson. New York: Oxford University Press.

Kramer, Randall A., Carel van Schaik, and Julie Johnson, eds. 1997. *Last Stand: Protected Areas and the Defense of Tropical Biodiversity.* New York: Oxford University Press.

Krebs, Charles J. 1972. *Ecology.* New York: Harper and Row.

Landauer, Walter. 1945. "Shall We Lose or Keep Our Plant and Animal Stocks?" *Science* 101, no. 2629: 497–499.

Lautenschlager, R. A. 1997. "Biodiversity Is Dead." *Wildlife Society Bulletin* 25, no. 3: 679–685.

Lehmann, V. W. 1938. "Some Values of Natural Areas." *Bird Lore* 40 (September): 310–14.

Leopold, Aldo. 1921. "The Wilderness and Its Place in Forest Recreational Policy." *Journal of Forestry* 19: 718–721.

———. 1933a. "The Conservation Ethic." *Journal of Forestry* 31: 634–643.

———. 1933b. *Game Management.* New York: Charles Scribner's Sons.

———. 1949. *A Sand County Almanac.* New York: Oxford University Press.

Leopold, A. Starker, S. A. Cain, C. M. Cottam, I. N. Gabrielson, and T. L. Kimball. 1964. "Predator and Rodent Control in the United States." In *Transactions of the Twenty-ninth North American Wildlife Conference.* Washington, DC: Wildlife Management Institute.

Lewontin, R. C. 1969. "The Meaning of Stability." In *Diversity and Stability in Ecological Systems,* Brookhaven Symposium of Biology 22, edited by G. M. Woodwell and H. H. Smith. Upton, NY: Brookhaven National Laboratories.

Lindeman, Raymond L. 1942. "The Trophic-Dynamic Aspect of Ecology." *Ecology* 23, no. 1: 399–417.

Loucks, Orie L. 1984. "Natural Diversity as a Scientific Concept in Resource

Management." In *Natural Diversity in Forest Ecosystems,* edited by James L. Cooley and June H. Cooley. Athens: Institute of Ecology, University of Georgia.

Lovejoy, Thomas E. 1980a. "Changes in Biological Diversity." In *The Global 2000 Report to the President (The Technical Report),* edited by G. O. Barney, 327–333. Harmondsworth, England: Penguin.

———. 1980b. Foreword to *Conservation Biology: An Evolutionary-Ecological Perspective,* edited by Michael E. Soulé and Bruce A. Wilcox. Sunderland, MA: Sinauer.

———. 1982. "Biological Diversity and Society." In *Proceedings of the U.S. Strategy Conference on Biological Diversity, November 16–18, 1981,* U.S. Agency for International Development. Washington, DC: Agency for International Development.

———. 1999. Interview with author by telephone. April 1.

Lund, Thomas A. 1980. *American Wildlife Law.* Berkeley and Los Angeles: University of California Press.

MacArthur, Robert H. 1955. "Fluctuations of Animal Populations and a Measure of Community Stability." *Ecology* 36, no.3: 533–536.

MacArthur, Robert H., and Joseph H. Connell. 1966. *The Biology of Populations.* New York: John Wiley and Sons.

MacArthur, Robert H., and Edward O. Wilson. 1967. *The Theory of Island Biogeography.* Princeton, NJ: Princeton University Press.

Mangelsdorf, Paul C. 1972. Introduction to *Plants in the Development of Modern Medicine,* edited by Tony Swain. Cambridge, MA: Harvard University Press.

Margalef, Ramon. 1963. "On Certain Unifying Principles in Ecology." *American Naturalist* 97, no. 897: 357–374.

———. 1969. "Diversity and Stability: A Practical Proposal and a Model of Interdependence." In *Diversity and Stability in Ecological Systems,* Brookhaven Symposium of Biology 22, edited by G. M. Woodwell and H. H. Smith. Upton, NY: Brookhaven National Laboratories.

Matthiessen, Peter. 1987. *Wildlife in America.* New York: Viking Penguin.

May, Robert. 1971. "Stability in Multi-species Community Models." *Mathematical Biosciences.* 12, no. 1: 59–79.

———. 1972. "Will a Large Complex System Be Stable?" *Nature* 238, no. 5364: 413–414.

———. 1973a. "Qualitative Stability in Model Ecosystems." *Ecology* 54, no. 3: 638–641.

———. 1973b. *Stability and Complexity in Model Ecosystems.* Princeton, NJ: Princeton University Press.

Mayr, Ernst. 1982. *The Growth of Biological Thought: Diversity, Evolution, and Inheritance.* Cambridge, MA: Belknap.

McCoy, J. J. 1970. *Saving Our Wildlife.* New York: Crowell-Collier.

McIntosh, Robert P. 1977. "Ecology since 1900." In *History of American Ecology,* edited by Frank N. Egerton. New York: Arno.

McManus, Roger. 2006. Interview with author by telephone. March 27.

McNaughton, S. J. 1977. "Diversity and Stability of Ecological Communities: A Comment on the Role of Empiricism in Ecology." *American Naturalist* 111, no. 979: 515–525.

Metcalf, Lee, and Keith Sebalius. 1978. "A Program for Preserving America's Natural Diversity." *Nature Conservancy News* 28, no. 1: 6–12.

Mighetto, Lisa. 1991. *Wild Animals and American Environmental Ethics.* Tucson: University of Arizona Press.

Milton, John P. 1966. "Retrospect." In *Future Environments of North America,* edited by F. Fraser Darling and John P. Milton. Garden City, NY: Natural History Press.

Moore, John A. 1963. *Heredity and Development.* New York: Oxford University Press.

Moore, N. W. 1969. "Experience with Pesticides and the Theory of Conservation." *Biological Conservation* 1, no. 3: 201–207.

Moyseenko, Helmut P. 1974. "A Comprehensive Natural Areas Program." *Nature Conservancy News* 24, no. 3: 19–20.

Mumford, Lewis. 1966. "Closing Statement." In *Future Environments of North America,* edited by F. Fraser Darling and John P. Milton. Garden City, NY: Natural History Press.

Myers, Norman. 1979. *The Sinking Ark: A New Look at the Problem of Disappearing Species.* New York: Pergamon.

———. 1980a. *Conversion of Tropical Moist Forests.* Washington, DC: Committee on Research Priorities in Tropical Biology of the National Research Council.

———. 1980b. "The Conversion of Tropical Moist Forests." *Environment* 22, no. 6: 6–13.

———. 1982. "Forest Refuges and Conservation in Africa." In *Biological Diversification in the Tropics,* edited by Ghillean T. Prance. New York: Columbia University Press.

———. 1983. *A Wealth of Wild Species: Storehouse for Human Welfare.* Boulder, CO: Westview.

Myers, Norman, and Edward S. Ayensu. 1983. "Reduction of Biological Diversity and Species Loss." *Ambio* 12, no. 2: 72–74.

National Research Council (NRC). 1972. *Genetic Vulnerability of Major Crops.* Washington, DC: National Academy of Sciences.

———. 1978. *Conservation of Germplasm Resources: An Imperative.* Washington, DC: National Academy of Sciences.

The Nature Conservancy (TNC). 1972a. "Mission Statement." *Nature Conservancy News* 22, no. 1: i.

———. 1972b. "Mission Statement." *Nature Conservancy News* 22, no. 3: i.

———. 1974. "Article II Amendments." *Nature Conservancy News* 24, no. 3: 16–18.

———. 1975a. *Preserving Our Natural Heritage: Federal Activities.* Washington, DC: Government Printing Office.

———. 1975b. *The Preservation of Natural Diversity.* Washington, DC: Nature Conservancy.

———. 1976. "Mission Statement." *Nature Conservancy News* 26, no. 1: i.

———. 1978. "Mission Statement." *Nature Conservancy News* 28, no. 1: i.

Nature Magazine. 1934. "The 'New Deal' for Waterfowl." *Nature Magazine* 24, no. 5: 197.

Norse, Elliott A. 1996. "A River That Flows to the Sea: The Marine Biological Diversity Movement." *Oceanography* 9, no. 1: 5–9.

———. 1999. Interview with author by telephone. March 31.

Norse, Elliott A., and Roger E. McManus. 1980. "Ecology and Living Resources: Biological Diversity." In *Environmental Quality 1980: The Eleventh Annual Report of the Council on Environmental Quality.* Washington, DC: Council on Environmental Quality.

Norse, Elliott A., Kenneth L. Rosenbaum, David S. Wilcove, Bruce A. Wilcox, William H. Romme, David W. Johnston, and Martha L. Stout. 1986. *Conserving Biological Diversity in Our National Forests.* Washington, DC: Wilderness Society.

Noss, Reed F. 1983. "A Regional Landscape Approach to Maintain Diversity." *BioScience* 33, no. 11: 700–706.

———. 1991. "From Endangered Species to Biodiversity." In *Balancing on the Brink of Extinction,* edited by Kathryn A. Kohm. Washington, DC: Island Press.

Odum, Eugene P. 1953. *Fundamentals of Ecology.* Philadelphia: W. B. Saunders.

———. 1962. "Relationships between Structure and Function in Ecosystems." *Japanese Journal of Ecology* 12: 108–118.

———. 1963. *Ecology.* New York: Holt, Rinehart, and Winston.

———. 1969. "The Strategy of Ecosystem Development." *Science* 164, no. 3877: 262–270.

———. 1984. "Diversity and the Forest Ecosystem." In *Natural Diversity in Forest Ecosystems,* edited by James L. Cooley and June H. Cooley. Athens: Institute of Ecology, University of Georgia.

Oelschlaeger, Max. 1991. *The Idea of Wilderness: From Prehistory to the Age of Ecology.* New Haven: Yale University Press.

Office of Technology Assessment (OTA). 1987. *Technologies to Maintain Biological Diversity, OTA-F-330.* Washington, DC: Government Printing Office.

Oldfield, Margery L. 1984. *The Value of Conserving Genetic Resources.* Washington, DC: Department of the Interior, National Park Service.

Owings, Loren C. 1976. *Environmental Values, 1860–1972: A Guide to Information Sources.* Detroit: Gale Research.

Pearson, G. A. 1922. "Preservation of Natural Areas in the National Forests." *Ecology* 3, no. 4: 284–287.

Pearson, T. Gilbert. 1918. "To Conserve Food in America, We Must Preserve the Wild Life." *Touchstone* 3: 257–260.

Perlman, Dan L., and Glenn Adelson. 1997. *Biodiversity: Exploring Values and Priorities in Conservation.* Malden, MA: Blackwell Science.

Peterson, R. Max. 1984. "Diversity Requirements in the National Forest Management Act." In *Natural Diversity in Forest Ecosystems,* edited by James L. Cooley and June H. Cooley. Athens: Institute of Ecology, University of Georgia.

Petulla, Joseph M. 1980a. *American Environmentalism: Values, Tactics, Priorities.* College Station: Texas A&M University Press.

———. 1980b. "Historic Values Affecting Wildlife in American Society." In *Wildlife Values,* edited by William W. Shaw and Ervin H. Zube. Tucson, AZ: Center for Assessment of Noncommodity Natural Resource Values.

———. 1987. "Evolution of the Valuation of Wildlife: A Historical Footnote." In *Valuing Wildlife: Economic and Social Perspectives,* edited by Daniel J. Decker and Gary R. Goff. Boulder, CO: Westview.

Pianka, Eric R. 1974. *Evolutionary Ecology.* New York: Harper and Row.

Piemeisel, R. L. 1940. "A Standard Experimental Vegetation Type." *Science* 92, no. 2383: 195–197.

Pimentel, David. 1982. "Biological Diversity and Environmental Quality." In *Proceedings of the U.S. Strategy Conference on Biological Diversity, November 16–18, 1981,* U.S. Agency for International Development. Washington, DC: Agency for International Development.

Pistorius, Robin, and Jereon van Wijk. 1999. *The Exploitation of Plant Genetic Information: Political Strategies in Crop Development.* New York: CABI.

Plucknett, Donald L., Nigel J. H. Smith, J. T. Williams, and N. Murthi Anishetty. 1987. *Gene Banks and the World's Food.* Princeton, NJ: Princeton University Press.

Preston, Frank W. 1969. "Diversity and Stability in the Biological World." In *Diversity and Stability in Ecological Systems,* Brookhaven Symposium of Biology 22, edited by G. M. Woodwell and H. H. Smith. Upton, NY: Brookhaven National Laboratories.

Proceedings of the North American Wildlife Conference Called by President Franklin D. Roosevelt. 1938. Washington, DC: Government Printing Office.

Radford, Albert E., Deborah Kay Strady Otte, Lee J. Otte, Jimmy R. Massey, and Paul D. Whitson. 1981. *Natural Heritage: Classification, Inventory, and Information.* Chapel Hill: University of North Carolina Press.

Reiger, John F. 1975. *American Sportsmen and the Origins of Conservation.* Norman: University of Oklahoma Press.

Risser, Paul G., and Kathy D. Cornelison. 1979. *Man and the Biosphere.* Norman: University of Oklahoma Press.

Rolston, Holmes, III. 1981. "Values in Nature." *Environmental Ethics* 3, no. 2: 113–128.

———. 1986. *Philosophy Gone Wild: Essays in Environmental Ethics.* Buffalo, NY: Prometheus.

———. 1988. *Environmental Ethics: Duties to and Values in the Natural World.* Philadelphia: Temple University Press.

———. 1994. *Conserving Natural Value.* New York: Columbia University Press.

Roosevelt, Theodore. 1915. "The Conservation of Wildlife." *Outlook* 109 (January 20): 159–162.

Roush, G. Jon. 1977. "Why Save Diversity?" *Nature Conservancy News* 27, no. 1: 9–12.

———. 1982. "On Saving Diversity." *Nature Conservancy News* 32, no. 1: 4–10.

Runte, Alfred. 1987. *National Parks: The American Experience.* Lincoln: University of Nebraska Press.

Rush, William. 1937. "What Are Wildlife Values?" *Nature Magazine* 30, no. 1: 40–43.

Salwasser, Hal. 1991. "In Search of an Ecosystem Approach to Endangered Species Conservation." In *Balancing on the Brink of Extinction,* edited by Kathryn A. Kohm. Washington, DC: Island Press.

Salwasser, Hal, Jack W. Thomas, and Fred Samson. 1984. "Applying the Diversity Concept to National Forest Management." In *Natural Diversity in Forest Ecosystems,* edited by James L. Cooley and June H. Cooley. Athens: Institute of Ecology, University of Georgia.

Schoenfeld, Clarence A., and John C. Hendee. 1978. *Wildlife Management in Wilderness.* Pacific Grove, CA: Boxwood.

Schonewald-Cox, Christine M., Steven M. Chambers, Bruce MacBryde, and W. Lawrence Thomas, eds. 1983. *Genetics and Conservation: A Reference for Managing Wild Animal and Plant Populations.* Menlo Park, CA: Benjamin/Cummings.

Schultes, Richard Evans. 1972. "The Future of Plants as Sources of New Biodynamic Compounds." In *Plants in the Development of Modern Medicine,* edited by Tony Swain. Cambridge, MA: Harvard University Press.

Scovell, E. L. 1938. "Overlook No Living Thing." *Recreation* 32 (August): 295–296.

Seigler, David S., ed. 1977. *Crop Resources: Proceedings of the 17th Annual Meeting of the Society for Economic Botany.* New York: Academic Press.

Sellars, Richard West. 1997. *Preserving Nature in the National Parks: A History.* New Haven: Yale University Press.

Shelford, Victor E. 1913. *Animal Communities in Temperate America.* Chicago: University of Chicago Press.

———, ed. 1926. *Naturalist's Guide to the Americas.* Baltimore: Williams and Wilkins.

———. 1933. "The Preservation of Natural Biotic Communities." *Ecology* 14, no. 2: 240–245.

———. 1943. "Twenty-five Year Effort at Saving Nature for Scientific Purposes." *Science* 98, no. 2543: 280–281.

———. 1963. *The Ecology of North America.* Chicago: University of Illinois Press.

Shen, Susan. 1987. "Biological Diversity and Public Policy." *BioScience* 37, no. 10: 709–712.

Simberloff, Daniel, and Lawrence Abele. 1976. "Island Biogeography Theory and Conservation in Practice." *Science* 191, no. 4224: 285–286.

Skutch, Alexander. 1948. "Earth and Man." *Audubon Magazine* 50 (November): 356–359.

Smith, Thomas B., and Robert K. Wayne, eds. 1996a. *Molecular Genetic Approaches in Conservation.* New York: Oxford University Press.

———. 1996b. Preface to *Molecular Genetic Approaches in Conservation,* edited by Thomas B. Smith and Robert K. Wayne. New York: Oxford University Press.

Soulé, Michael E. 1985. "What Is Conservation Biology?" *BioScience* 35, no. 11: 727–734.

———. 1987. "News of the Society." *Conservation Biology* 1, no. 1: 4.

Soulé, Michael E., and Bruce A. Wilcox, eds. 1980a. *Conservation Biology: An Evolutionary-Ecological Perspective.* Sunderland, MA: Sinauer.

———. 1980b. Introduction to *Conservation Biology: An Evolutionary-Ecological Perspective,* edited by Michael E. Soulé and Bruce A. Wilcox. Sunderland, MA: Sinauer.

Southwick, Charles H. 1972. *Ecology and the Quality of Our Environment.* New York: D. Van Norstrand.

Steinhoff, Harold W. 1980. "Analysis of Major Conceptual Systems for Understanding and Measuring Wildlife Values." In *Wildlife Values,* edited by William W. Shaw and Ervin H. Zube. Tucson, AZ: Center for Assessment of Noncommodity Natural Resource Values.

Steinhoff, Harold W., Richard G. Walsh, Tony J. Peterle, and Joseph M. Petulla. 1987. "Evolution of the Valuation of Wildlife." In *Valuing Wildlife: Economic and Social Perspectives,* edited by Daniel J. Decker and Gary R. Goff. Boulder, CO: Westview.

Stoddard, Herbert. 1931. *The Bobwhite Quail: Its Habits, Preservation, and Increase.* New York: Charles Scribner's Sons.

Stone, Carol Beall. 1984. "From Forests to Fields to Food Webs: The Environment in History and Biology Textbooks, 1905–1975." Ph.D. diss., Stanford University.

Stone, Peter. 1973. *Did We Save the Earth at Stockholm?* London: Earth Island.

Sturtevant, A. H. 1965. *A History of Genetics.* New York: Harper and Row.

Sumner, F. B. 1921. "The Responsibility of the Biologist in the Matter of Preserving Natural Conditions." *Science* 54, no. 1385: 39–43.

Takacs, David. 1996. *The Idea of Biodiversity: Philosophies of Paradise.* Baltimore: Johns Hopkins University Press.

Tangley, Laura. 1985. "A New Plan to Conserve the Earth's Biota." *BioScience* 35, no. 6: 334–336+.

———. 1986. "Biological Diversity Goes Public." *BioScience* 36, no. 11: 708–711+.

Tansley, Arthur G. 1935. "The Use and Abuse of Vegetational Concepts and Terms." *Ecology* 16, no. 3: 284–307.

Taylor, Norman. 1966. *Plant Drugs That Changed the World.* New York: Dodd, Mead.

Terborgh, John. 1976. "Island Biogeography and Conservation: Strategy and Limitations." *Science* 193, no. 4257: 1029–1030.

Thomas, Keith. 1983. *Man and the Natural World: A History of the Modern Sensibility.* New York: Pantheon.

Trefethen, James B. 1975. *An American Crusade for Wildlife.* New York: Winchester.

United Nations Educational, Scientific, and Cultural Organization (UNESCO). 1970. *Use and Conservation of the Biosphere.* Paris: UNESCO.

———. 1973. *Programme on Man and the Biosphere: Expert Panel on Project 8; Conservation of Natural Areas and of the Genetic Material They Contain, Final Report.* Morges, Switzerland: UNESCO.

———. 1974. *Programme on Man and the Biosphere: Task Force on Criteria and Guidelines for Choice and Establishment of Biosphere Reserves.* Paris: UNESCO.

U.S. Agency for International Development (USAID). 1982. *Proceedings of the U.S. Strategy Conference on Biological Diversity, November 16–18, 1981.* Washington, DC: Agency for International Development.

———. 1985. *U.S. Strategy on the Conservation of Biological Diversity: An Interagency Task Force Report to Congress.* Washington, DC: Agency for International Development.

U.S. Congress. 1973. *Endangered Species Act of 1973.* 93rd Cong., 1st sess. Public Law 93-205. Washington, DC: Government Printing Office.

———. House. Committee on Foreign Affairs. Subcommittee on Human Rights and International Organizations. 1985. *U.S. Policy on Biological Diversity.* 99th Cong., 1st sess., June 6. Washington, DC: Government Printing Office.

Van Name, Willard G. 1919. "Zoological Aims and Opportunities." *Science* 50, no. 1282: 81–84.

———. 1941. "Need for the Preservation of Natural Areas Exemplifying Vegetation Types." *Science* 93, no. 2418: 423.

Vavilov, N. I. 1926. *Studies on the Origin of Cultivated Plants.* Leningrad: Institute of Applied Botany and Plant Breeding.

Wallace, Henry A. 1934. "Give Research a Chance." *Country Gentlemen* 104, no. 9: 5–6+.

———. 1961. "Origin and Utilization of Germ Plasm in the United States." In *Germ Plasm Resources,* edited by Ralph E. Hodgson. Washington, DC: American Association for the Advancement of Science.

Warren, Louis. 1997. *The Hunter's Game: Poachers and Conservationists in Twentieth-Century America.* New Haven: Yale University Press.

Wenzel, Lauren. 1985. "Congress Considers Biological Diversity." *BioScience* 35, no. 11: 696.

Whitcomb, Robert, James Lynch, Paul Opler, and Chandler Robbins. 1976. "Island Biogeography and Conservation: Strategy and Limitations." *Science* 193, no. 4257: 1030–1032.

Whitehouse, H. L. K. 1973. *Towards an Understanding of the Mechanism of Heredity.* New York: St. Martin's.

Whittaker, Robert H. 1975. *Communities and Ecosystems.* 2nd ed. New York: Macmillan.

Wieting, Hardy, Jr. 1976. "The Protected Natural Areas Project." *Nature Conservancy News* 26, no. 2: 31–35.

Wilcox, Bruce A. 1984. "Concepts in Conservation Biology: Application to the Management of Biological Diversity." In *Natural Diversity in Forest Ecosystems,* edited by James L. Cooley and June H. Cooley. Athens: Institute of Ecology, University of Georgia.

Wilkes, Garrison. 1983. "Current Status of Crop Plant Germplasm." *CRC Critical Reviews in Plant Sciences* 1, no. 2: 133–181.

Williams, J. T. 1984. "A Decade of Crop Genetic Resources Research." In *Crop Genetic Resources: Conservation and Evaluation,* edited by J. H. W. Holden and J. T. Williams. London: George Allen and Unwin.

Wilshusen, Peter, Steven R. Brechin, Crystal L. Fortwangler, and Patrick C. West. 2002. "Reinventing the Square Wheel: Critique of a Resurgent 'Protection Paradigm' in International Biodiversity Conservation." *Society and Natural Resources* 15, no. 1: 17–40.

Wilson, Edward O. 1984. "Million-Year Histories: Species Diversity as an Ethical Goal." *Wilderness* 47 (Summer): 12–17.

———. 1985. "The Biological Diversity Crisis: A Challenge to Science." *Issues in Science and Technology* 2, no. 1: 20–29.

———, ed. 1988a. *Biodiversity.* Washington, DC: National Academy Press.

———. 1988b. "The Current State of Biological Diversity." In *Biodiversity,* edited by Edward O. Wilson. Washington, DC: National Academy Press.

———. 1992. *The Diversity of Life.* New York: W. W. Norton.

Witt, Steven C. 1985. *Biotechnology and Genetic Diversity.* San Francisco: California Agricultural Lands Project.

Woodwell, G. M., and H. H. Smith. 1969. Preface to *Diversity and Stability in Ecological Systems,* Brookhaven Symposium of Biology 22, edited by G. M. Woodwell and H. H. Smith. Upton, NY: Brookhaven National Laboratories.

Worster, Donald. 1994. *Nature's Economy: A History of Ecological Ideas.* New York: Cambridge University Press.

Worthington, E. B., ed. 1975a. *The Evolution of IBP.* Cambridge: Cambridge University Press.

———. 1975b. "Substance of the Programme." In *The Evolution of IBP. Edited by E. B. Worthington.* Cambridge: Cambridge University Press.

Yaffee, Steven Lewis. 1982. *Prohibitive Policy: Implementing the Federal Endangered Species Act.* Cambridge, MA: MIT Press.

Zohary, Daniel. 1970. "Centers of Diversity and Centers of Origin." In *Genetic Resources in Plants: Their Exploration and Conservation,* edited by Otto H. Frankel and Erna Bennett. Philadelphia: F. A. Davis.

Index

808.3
BAU

c.1

Bauer, Marion Dane.

Our stories.

34880030002161

$16.0

808.3
BAU
c.1

Bauer, Marion D

Our Stories

$16.00

34880300

5169

348803002161

DATE DUE	BORROWER'S NAME	ROOM NUMBER